Creating

NUMBERS
and
NUMBER SYSTEMS

Edward G. Fleming

You can buy this book on Amazon.com

Last Edited on June 20, 2024

To my grandkids, and their kids, and so on and so on…
may you enjoy mathematics as much as I do.

The views and opinions in this book are my own and not necessarily those of the Mathematical Community.

Edward G. Fleming

Contents

Contents

Introduction

What are numbers? What are numerals? Are they the same? How precise are they? This book explores these questions in detail and develops a method of calculation with numbers that are not precise. Examples are given for each new application. The book concludes with the discussion of a new technique for the error analysis.

In Part I of this book, we will look at what is meant by a precise number and their properties. The following is a list of the systems of precise numbers that are discussed in this book:

- Natural Numbers
- Integers
- Rational Numbers
- Real Numbers
- Complex Numbers
- Another 2-Dimensional Number System (Cartesian Coordinate System)
- Our 3-Dimensional Number System -
  Modeled on Our Own 3-Dimensional Physical World
- Another 3-Dimensional Number System
- A possible 4-Dimensional Number System

The second part of the book is devoted to numbers that are close to precise numbers, but aren't themselves, precise. We will call these numbers, *imprecise numbers*. Each imprecise number can be represented by a *collection* of precise numbers. In a *set* of numbers, say, $A = \{a_1, a_2, a_3, \ldots, a_n\}$, no two elements are the same. But a collection will be defined as a grouping of numbers in which any number may be repeated any number of times. We will denote a collection by using the double brackets, $[\![\ ]\!]$. A collection X that contains all the numbers, $x_1, x_2, x_3, \ldots, x_m$, will be written as $X = [\![x_1, x_2, x_3, \ldots, x_m]\!]$. Each entry in the collection has an equal probability of being an acceptable value for X, so that the more a given number is repeated, the better will be its chance of describing the imprecise number.

A collection of numbers that describes an imprecise number may be very large, so we need to develop techniques for summarizing their properties. The first method for working with imprecise numbers that we will explore involves simplifying them by finding the means and standard deviations of the collections representing them. The following generalizations are proven in this book.

Let $X = [\![x_1, x_2, x_3, \ldots, x_m]\!]$ and $Y = [\![y_1, y_2, y_3, \ldots, y_n]\!]$, where x_i and y_j are real numbers and let k be a real number, then I will define $f(X) = [\![f(x_1), f(x_2), f(x_3), \ldots, f(x_m)]\!]$ and

$$f(X,Y) = \begin{Vmatrix} f(x_1,y_1), f(x_1,y_2), f(x_1,y_3), \cdots, f(x_1,y_n), \\ f(x_2,y_1), f(x_2,y_2), f(x_2,y_3), \cdots, f(x_2,y_n), \\ f(x_3,y_1), f(x_3,y_2), f(x_3,y_3), \cdots, f(x_3,y_n), \\ \vdots \quad\quad \vdots \quad\quad \vdots \quad\quad \ddots \quad\quad \vdots \\ f(x_m,y_1), f(x_m,y_2), f(x_m,y_3), \cdots, f(x_m,y_n) \end{Vmatrix}.$$

$$f(X) = X + k$$
$$\mu_{X+k} = \mu_X + k \text{ and } \sigma_{X+k} = \sigma_X$$

$$f(X) = X - k$$
$$\mu_{X-k} = \mu_X - k \text{ and } \sigma_{X-k} = \sigma_X$$

$$f(X,Y) = X + Y$$
$$\mu_{X+Y} = \mu_X + \mu_Y \text{ and } \sigma_{X+Y} = \sqrt{\sigma_X^2 + \sigma_Y^2}$$

$$f(X,Y) = X - Y$$
$$\mu_{X-Y} = \mu_X - \mu_Y \text{ and } \sigma_{X-Y} = \sqrt{\sigma_X^2 + \sigma_Y^2}$$

$$f(X) = kX$$
$$\mu_{kX} = k\mu_X \text{ and } \sigma_{kX} = |k|\sigma_X$$

$$f(X,Y) = XY$$
$$\mu_{XY} = \mu_X \mu_Y \text{ and } \sigma_{XY} = \sqrt{\mu_X^2 \sigma_Y^2 + \mu_Y^2 \sigma_X^2 + \sigma_X^2 \sigma_Y^2}$$

$f(X,Y) = \dfrac{X}{Y}$, this will not be defined by the above method of finding $f(X,Y)$,

$f(X,Y) = \dfrac{X}{Y}$ will be defined as a collection where $\mu_{Y\left(\frac{X}{Y}\right)} = \mu_X$ and $\sigma_{Y\left(\frac{X}{Y}\right)} = \sigma_X$.

$$\mu_{\frac{X}{Y}} = \frac{\mu_X}{\mu_Y} \text{ and } \sigma_{\frac{X}{Y}} = \sqrt{\frac{\sigma_X^2 - \left(\dfrac{\mu_X}{\mu_Y}\right)^2 \sigma_Y^2}{\mu_Y^2 + \sigma_y^2}}$$

The second approach that we will explore for working with imprecise numbers involves determining their acceptable maximum and minimum end point values. Once we have found these values, we will discuss techniques for calculating the resultants after applying different functions and operations.

CREATING

NUMBERS

and

Number Systems

Part I

The Precise Numbers

Chapter 1

Precise Numbers

1.1 Numbers and Their Precision

A *number* is an abstract concept with some properties, while a *numeral* is a symbol that represents a number. The problem with numerals is that the numeral 1, for example, may represent anything from a precise value of one (without any margin for error), to something that is close to one, depending on the context in which it is used.

In this book, we will be focusing very closely on the numbers and their properties. We are not interested in numerals. We will consider two different kinds of numbers: *precise* numbers and *imprecise* numbers. The difference between these two kinds of numbers lies in their *degree of precision*. In a precise number, there is no room for *error*, whereas an imprecise number can include some room for error if its value is reasonably close to the stated value. In this chapter, the focus will be only on precise numbers.

1.2 In the Beginning

A Fictional Creation Story

In the beginning there was *nothing, absolutely nothing*. There was nothing everywhere. All there was… was nothing. I thought and thought. It was hard work thinking about nothing. In fact, it was PURE TORTURE having nothing to think about. All this thinking about nothing wore me out and I fell asleep.

When I woke up, after a long time of thinking about "the nothing", a thought came to my mind. I will call all this nothingness "Nothing." Creating the name, "Nothing" was very hard work, and I fell asleep, again.

After I woke up, I thought and thought some more. There was nothing everywhere and there was the name, "Nothing". The name, "Nothing" was not the nothingness itself, but something else; it was the label of the nothing that named, "Nothing". But "Nothing" was also nothing. This confused me. How could the name or label, "Nothing", be nothing and not be nothing all at the same time? Looking more closely at the name, "Nothing" I realized that it came from "the nothingness." The nothingness was there first and after that came the name or label, "Nothing". So, the name or label, "Nothing" was like another nothing. I gave the name or label "Nothing" another name or label, "Nothing Next." All of this was hard work and confusing. Looking back on it there now was the nothing, the name "Nothing", and the label of the name "Nothing", that I called "Nothing Next." This too was hard work and very confusing, so I fell asleep.

The next day after I woke, I thought and thought some more. Naming the nothings was hard work, but enjoyable. Recapping, I now had the original nothingness, its name "Nothing", and the label of the name "Nothing", as "Nothing Next." Why stop there, why not name "Nothing Next?" I had enjoyed naming things before, so I thought I would enjoy naming them again. So, I gave the label "Nothing Next" the name, "Nothing Too." But this was too confusing and not enjoyable at all. There were too many nothings. I thought and thought. Then a thought came to me: I could rename the names I have made form, "Nothing" to "Zero," and "Nothing Next" to "One". I will change "Nothing Too" to "Two." All of this was very hard work, but enjoyable, now that I had the original nothing, the name, "Zero", "One", and "Two". And fell asleep.

The next day after I awoke, I thought and thought some more. There was the original nothingness, and the names "Zero", "One", and "Two". I could continue this process forever, renaming each name of the preceding name of nothingness as it arose, and I did until I fell asleep.

The next day, being a mathematician, I needed to generalize what I had done. Starting with the original nothing, then it's label or name "Zero", the next successor I got was "One", I got its successor "Two", and so on, and so on. But being a mathematician, I didn't like writing all this out, so I shortened it from "Zero" to 0, "One" to 1, "Two" to 2, and so on and so on. Now I had the label or the name of the original nothing as 0, and it's successor as equaling 1, the successor of 1 equaling 2, and so on and so on. But that still wasn't short enough for me, so I shorten the word successor to the letter S and if I want the successor of a number, x, I would simply write it as $S(x)$. So, I now would write the following: $S(0)=1$, $S(1)=2$, $S(2)=3$, and so on and so on. I could also write, $S(0)=1$, $S(S(0))=2$, and $S(S(S(0)))=3$ and so on and so on. I liked shorting so much I need to shorten the original nothing to 0. So, now I could wite $S(0)=1$, $S(S(0))=2$, $S(S(S(0)))=3$, and so on and so on. I would call the set of labels $\{0,1,2,3,...\}$ numbers. After working hard, I then fell asleep.

On the seventh day I just rested. All that thinking about nothing had wore me out. It was fun recapping all that had happened those last six days. I was tired, but felt good. So, I said "This is good." So, I rested.

Awaking early the next day. I asked myself what would happen if I combined two numbers. Let me put 1 with 2 and see what I get. Let's see $1=S(0)$ and $2=S(S(0))$ so combining 1 and 2, I would get $S(S(S(0)))$ that was equal to 3. But there was a problem, combining was too long a word. So, I shortened it to add. If I add 1 and 2 that would get 3. But that wasn't shorten enough for me, so I made a plus sign, $+$, The plus sign meant we were combining numbers in this way. So now, $1+2=3$. And another day ended.

I awoke with a realization, if m and n has a one-to-one correscondence of successor of 0 that m and n are equal. Examaple:

$$m = S\Big(S\big(S\big(S\big(S\big(S\big(S(S(0))\big)\big)\big)\big)\big)\Big)$$

$$n = S\Big(S\big(S\big(S\big(S\big(S\big(S(S(0))\big)\big)\big)\big)\big)\Big)$$

then

$$m = n$$

I could prove the $a+b=b+a$ and that $(a+b)+c = a+(b+c)$, they both don't destroy the one-to-one correlation of successor of 0, by moving sets of successors around. After thinking about this all day, I fell asleep.

The next day after I awoke I decided to summaried what I had done, there was the original nothing, the nothings that were labels of the original nothing that I called Fleming's numbers.

1. Number 0 belonged to the set of numbers.
2. The number 0 was not the successor of any number.
3. If m was a number, then its successor $S(m)$ was a number.
4. If two numbers m and n, had a one-to-one correlation of successor of 0 between them, then $m=n$

I called these properties *"Fleming's Axioms"* or *"Fleming's Postulates of Numbers."*

Theorem 1.2.1:

Let $m = a + b$ and $n = b + a$, where a and b are Fleming's numbers as decided earlier. Then $m = n$ for all a and b which are a Fleming's number.

Prove:

Let $m = a + b$ and $n = b + a$, where a and b are Fleming's numbers.

Then a can be written as $a = S_1\left(S_2\left(S_3\left(\cdots S_a(0)\right)\right)\right)$

And b can be written as $b = S_1\left(S_2\left(S_3\left(\cdots S_b(0)\right)\right)\right)$

Therefore $m = S_1\left(S_2\left(S_3\left(\cdots S_a\left(S_1\left(S_2\left(S_3\left(\cdots S_b(0)\right)\right)\right)\right)\right)\right)\right)$

And $n = S_1\left(S_2\left(S_3\left(\cdots S_b\left(S_1\left(S_2\left(S_3\left(\cdots S_a(0)\right)\right)\right)\right)\right)\right)\right)$

There is a one-to-one correlation of successor of 0 between m and n as follows:

$$m = S_1\left(S_2\left(S_3\left(\cdots S_a\left(S_1\left(S_2\left(S_3\left(\cdots S_b(0)\right)\right)\right)\right)\right)\right)\right)$$

$$n = S_1\left(S_2\left(S_3\left(\cdots S_b\left(S_1\left(S_2\left(S_3\left(\cdots S_a(0)\right)\right)\right)\right)\right)\right)\right)$$

Therefore $m = n$ or you could say $a + b = b + a$
End of Proof

Theorem 1.2.2:

Let $m = (a + b) + c$ and $n = a + (b + c)$, where a, b, and c are Fleming's numbers as decided earlier.
Then $m = n$ for all a, b, and c which are a Fleming's number.

Proof:

Let $m = (a + b) + c$ and $n = a + (b + c)$, where a, b, and c are Fleming's numbers

Then a can be written as $a = S_1\left(S_2\left(S_3\left(\cdots S_a(0)\right)\right)\right)$

And b can be written as $b = S_1\left(S_2\left(S_3\left(\cdots S_b(0)\right)\right)\right)$

And c can be written as $c = S_1\left(S_2\left(S_3\left(\cdots S_c(0)\right)\right)\right)$

Therefore $m = S_1\left(S_2\left(S_3\left(\cdots S_a\left(S_1\left(S_2\left(S_3\left(\cdots S_b\left(S_1\left(S_2\left(S_3\left(\cdots S_c(0)\right)\right)\right)\right)\right)\right)\right)\right)\right)\right)\right)$

And $n = S_1\left(S_2\left(S_3\left(\cdots S_b\left(S_1\left(S_2\left(S_3\left(\cdots S_c\left(S_1\left(S_2\left(S_3\left(\cdots S_a(0)\right)\right)\right)\right)\right)\right)\right)\right)\right)\right)\right)$

There is a one-to-one correlation of successor of 0 between m and n as follows:

$$m = S_1\left(S_2\left(S_3\left(\cdots S_a\left(S_1\left(S_2\left(S_3\left(\cdots S_b\left(S_1\left(S_2\left(S_3\left(\cdots S_c(0)\right)\right)\right)\right)\right)\right)\right)\right)\right)\right)\right)$$

$$n = S_1\left(S_2\left(S_3\left(\cdots S_b\left(S_1\left(S_2\left(S_3\left(\cdots S_c\left(S_1\left(S_2\left(S_3\left(\cdots S_a(0)\right)\right)\right)\right)\right)\right)\right)\right)\right)\right)\right)$$

Therefore $m = n$ or you could say $(a+b)+c = a+(b+c)$

End of Proof

1.3 Natural Numbers

The set of *natural numbers*, $\mathbb{N} = \{0, 1, 2, 3, \ldots\}$. (A note to the reader: zero is not always included in natural numbers, but we will include it in the set of natural numbers in this book. Historically, some mathematicians were reluctant to accept zero as a number).

Let m, n, and p be natural numbers. Then there are natural numbers r and s such that

$m + n = r$	Closure under Addition
$m + n = n + m$	Commutative Law of Addition
$m + (n + p) = (m + n) + p$	Associative Law of Addition
If $m + n = m + p$, then $n = p$	Cancellation Law of Addition
$m + 0 = m$	Identity Element for Addition
$m \cdot n = s$	Closure under Multiplication
$m \cdot n = n \cdot m$	Commutative Law of Multiplication
$m \cdot (n \cdot p) = (m \cdot n) \cdot p$	Associative Law of Multiplication
If $m \neq 0$ and $m \cdot n = m \cdot p$, then $n = p$	Cancellation Law of Multiplication
$m \cdot 1 = m$	Identity Element for Multiplication
$(m + n) \cdot p = m \cdot p + n \cdot p$	Distributive Law

Either $m \leq n$ or $m \geq n$	Totality
If $m \leq n$ and $m \geq n$, then $m = n$	Anti-symmetry
If $m \leq n$ and $n \leq p$, then $m \leq p$	Transitivity

If $m \leq n$, then $m + p \leq n + p$
If $m \leq n$, then $mp \leq np$

1.4 Integers

From the natural numbers we can construct the set of all the *integers,* $\mathbb{Z} = \{\ldots, -3, -2, -1, 0, 1, 2, 3, \ldots\}$ by adding in the *negative numbers,* which are the opposites of the positive numbers. This is an idea that exists in one's mind, but it does not always have a concrete realization. For example, what is a negative one orange? We can talk about owing an orange to someone, but that is not exactly the same as having a negative orange. If positive is owning, then negative is the opposite of owning which is owing. If positive is owing, then negative is owning. If positive is up, then negative is down. If positive is down, then negative is up, etc. Positive is the opposite of negative and visi-versa.

When we are working with integers, we need to add the following properties to our list of the properties of the natural numbers:

Let m and n be integers. Then there is an integer p such that

$$m - n = p \qquad \text{Closure under subtraction}$$

For each integer m, there is an integer n, such that:

$$m + n = n + m = 0$$

The opposite of the integer m is called the Additive Inverse and is denoted by "$-m$".

The properties of integers are as follows:

Let m, n, and p be integers. Then there are integers r, s, and t such that

$m + n = r$	Closure under Addition
$m + n = n + m$	Commutative Law of Addition
$m + (n + p) = (m + n) + p$	Associative Law of Addition
If $m + n = m + p$, then $n = p$	Cancellation Law of Addition
$m + 0 = m$	Identity Element for Addition
There is a unique integer $-m$ such that $m + (-m) = 0$	Additive Inverse
$m \cdot n = s$	Closure under Multiplication
$m \cdot n = n \cdot m$	Commutative Law of Multiplication
$m \cdot (n \cdot p) = (m \cdot n) \cdot p$	Associative Law of Multiplication
If $m \neq 0$ and $m \cdot n = m \cdot p$, then $n = p$	Cancellation Law of Multiplication
$m \cdot 1 = m$	Identity Element for Multiplication
$(m + n) \cdot p = m \cdot p + n \cdot p$	Distributive Law
$m - n = t$	Closure under subtraction
Either $m \leq n$ or $m \geq n$	Totality
If $m \leq n$ and $m \geq n$, then $m = n$	Anti-symmetry
If $m \leq n$ and $n \leq p$, then $m \leq p$	Transitivity
If $m \leq n$, then $m + p \leq n + p$	
If $m \leq n$ and $p \geq 0$, then $mp \leq np$	
If $m \leq n$ and $p \leq 0$, then $mp \geq np$	

8

1.5 Rational Numbers

Now we can construct the set of all the *rational numbers,* $\mathbb{Q}$, by introducing the division operation on the set of integers. Of course, we have to remember that division by zero has no meaning. The rational numbers, $\mathbb{Q}$, are the numbers of the form m/n where m and n are integers and $n \neq 0$.

When we're working with rational numbers, all the properties of the integers are satisfied, and we can add these additional properties to our set of properties.

Let m and n be rational numbers and suppose that $n \neq 0$. Then there is a rational number p such that

$$p = \frac{m}{n} \qquad \text{Closure under Division}$$

If m is a rational number, then there exists an integer n such that $m \cdot n$ is an integer.

Below is the list of some of the properties of rational numbers:

Let m, n, and p be rational numbers. Then there are rational numbers, r, s, t, and u such that:

$m + n = r$	Closure under Addition
$m + n = n + m$	Commutative Law of Addition
$m + (n + p) = (m + n) + p$	Associative Law of Addition
If $m + n = m + p$, then $n = p$	Cancellation Law of Addition
$m + 0 = m$	Identity Element for Addition
There is a unique rational number $-m$ such that $m + (-m) = 0$	Additive Inverse
$m \cdot n = s$	Closure under Multiplication
$m \cdot n = n \cdot m$	Commutative Law of Multiplication
$m \cdot (n \cdot p) = (m \cdot n) \cdot p$	Associative Law of Multiplication
If $m \neq 0$ and $m \cdot n = m \cdot p$, then $n = p$	Cancellation Law of Multiplication
$m \cdot 1 = m$	Identity Element for Multiplication
$(m + n) \cdot p = m \cdot p + n \cdot p$	Distributive Law
$m - n = t$	Closure under subtraction
If $n \neq 0$, then $\dfrac{m}{n} = u$	Closure under Division
If $m \neq 0$, then $m\left(\dfrac{1}{m}\right) = 1$	Multiplicative Inverse $p = \dfrac{m}{n}$
Either $m \leq n$ or $m \geq n$	Totality
If $m \leq n$ and $m \geq n$, then $m = n$	Anti-symmetry
If $m \leq n$ and $n \leq p$, then $m \leq p$	Transitivity
If $m \leq n$, then $m + p \leq n + p$	
If $m \leq n$ and $p \geq 0$, then $mp \leq np$	
If $m \leq n$ and $p \leq 0$, then $mp \geq np$	

1.6 Real Numbers

There is one problem with the set of rational numbers. There are many numbers that are not a part of this set. Hence, we say that the set of rational numbers is not *complete*. For example, the square root of two is a precise number which, when multiplied by itself, gives the precise number two, that is. Let's write the square root of two as $\sqrt{2}$. It is easy to see that the precise number exists by the following example:

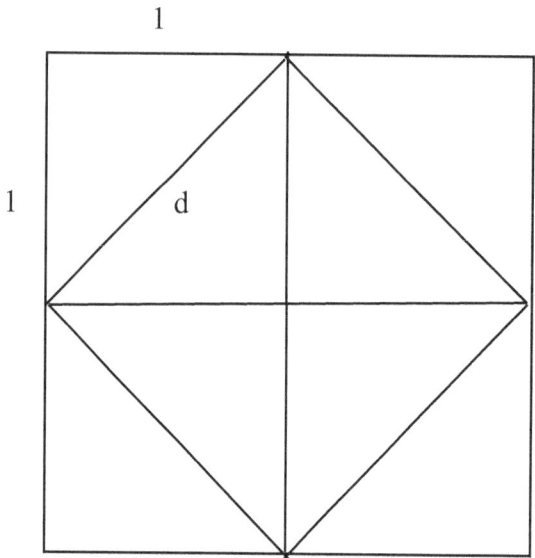

The top left box is a square with all four sides equaling 1, so that it has an area of 1. It is bisected by a diagonal line of length d, which then divides the square into two triangles with each having an area of one half. The four inner triangles make a square diamond shaped box with sides of length d. Adding the area of the four inner triangles we get a total area of 2. Since the inner diamond shape box is a square, the formula for the area is $A = d^2$ and its area is 2, Therefore $d^2 = 2$ or $d = \sqrt{2}$.

All the numbers that are not rational, such as $\sqrt{2}$, are called *irrational numbers*. The set of rational numbers, when combined with the set of irrational numbers, makes the set of *real numbers*. The set of real numbers has all of the properties of the set of rational numbers that we talked about in the earlier sections.

We can prove that $\sqrt{2}$ is not rational. But first, we need to prove the following preliminary lemma.

Lemma 1.6.1

If the square of an integer is odd, then that integer is odd. If the square of an integer is even, then that integer is even.

Proof:

We first prove that if the square of an integer is even, then that integer is also even. Let n be an integer and suppose that n^2 is even. We need to show that n is also even. Suppose that n is odd. Then there is an integer k such that $n = 2 \cdot k + 1$. But then,

$$
\begin{aligned}
n^2 &= (2k+1)^2 \\
&= (2 \cdot k)^2 + 2 \cdot (2 \cdot k) \cdot 1 + 1^2 \\
&= (2^2) \cdot (k^2) + 2 \cdot 2 \cdot k + 1 \\
&= 2 \cdot (2 \cdot k^2 + 2 \cdot k) + 1.
\end{aligned}
$$

Let $l = \left(2 \cdot k^2 + 2 \cdot k\right)$. Then l is an integer since integers are closed under addition and multiplication and we may write $n^2 = 2 \cdot l + 1$, which is odd. This contradicts our original assumption that n^2 was even. Therefore, n must be even.

We now must prove that if the square of an integer is odd, then the integer is also odd. Let n be an integer and suppose that n^2 is odd. We need to prove that n is also odd. Suppose that n is even. Then there's an integer k such that $n = 2k$. We calculate

$$
\begin{aligned}
n^2 &= (2k)^2 \\
&= 2^2 \cdot k^2 \\
&= 2 \cdot (2 \cdot k^2).
\end{aligned}
$$

If we write $l = 2 \cdot k^2$, then l is an integer and $n^2 = 2 \cdot l$, which is even. This contradicts our assumption that n^2 was even. Therefore, n must be odd.

End of proof.

Theorem 1.6.1:

The square root of two is irrational.

Proof :

We can prove that $\sqrt{2}$ is an irrational number by contradiction. That is we start with the assumption that $\sqrt{2}$ is irrational and show that this leads us to something that cannot possibly be true.

Suppose that $\sqrt{2}$ is rational. Then there are integers m and n with $n \neq 0$ such that

$$\sqrt{2} = \frac{m}{n}.$$

We may assume that $\frac{m}{n}$ is already in its most simplified form. Now, squaring both sides gives

$$\left(\sqrt{2}\right)^2 = \left(\frac{m}{n}\right)^2,$$

so that $2 = \frac{m^2}{n^2}$. Multiplying both sides by n^2 gives

$$2 \cdot \left(n^2\right) = \left(\frac{m^2}{n^2}\right) \cdot \left(n^2\right) = m^2.$$

Since n^2 is an integer, the number $2 \cdot n^2$ is even, and so m^2 is even. Therefore, m is even by Lemma 1.6.1.

Because m is even, there exists an integer p such that $m = 2 \cdot p$. Then

$$2n^2 = m^2 = (2 \cdot p)^2 = 2 \cdot (2 \cdot p^2).$$

Dividing both sides by 2, we can see that $n^2 = 2 \cdot p^2$. Since p^2 is an integer, this implies that n^2 is even. Therefore, n is even by Lemma 1.6.1. This means that we can find an integer q such that $n = 2 \cdot q$. But then, m and n share a common factor of 2, which contradicts our assumption that the fraction $\frac{m}{n}$ was in its most simplified form.

Therefore, $\sqrt{2}$ cannot be a rational number.

End of proof.

Below is the list of some of the properties of real numbers:

Let m, n, and p be real numbers. Then there are real numbers r, s, t, and u such that

$m + n = r$	Closure under Addition
$m + n = n + m$	Commutative Law of Addition
$m + (n + p) = (m + n) + p$	Associative Law of Addition
If $m + n = m + p$, then $n = p$	Cancellation Law of Addition
$m + 0 = m$	Identity Element for Addition
There is a unique real number $-m$ such that $m + (-m) = 0$	Additive Inverse
$m \cdot n = s$	Closure under Multiplication
$m \cdot n = n \cdot m$	Commutative Law of Multiplication
$m \cdot (n \cdot p) = (m \cdot n) \cdot p$	Associative Law of Multiplication
If $m \neq 0$ and $m \cdot n = m \cdot p$, then $n = p$	Cancellation Law of Multiplication
$m \cdot 1 = m$	Identity Element for Multiplication
$(m + n) \cdot p = m \cdot p + n \cdot p$	Distributive Law
$m - n = t$	Closure under subtraction
If $n \neq 0$, then $\dfrac{m}{n} = u$	Closure under Division
If $m \neq 0$, then $m\left(\dfrac{1}{m}\right) = 1$	Multiplicative Inverse $p = \dfrac{m}{n}$
Either $m \leq n$ or $m \geq n$	Totality
If $m \leq n$ and $m \geq n$, then $m = n$	Anti-symmetry
If $m \leq n$ and $n \leq p$, then $m \leq p$	Transitivity
If $m \leq n$, then $m + p \leq n + p$	
If $m \leq n$ and $p \geq 0$, then $mp \leq np$	
If $m \leq n$ and $p \leq 0$, then $mp \geq np$	

1.7 Complex Numbers

Not all numbers are real numbers. Real numbers provide solutions to a lot of equations, but not all. For example, there is no real solution to the equation $x \cdot x + 1 = 0$ or $x^2 + 1 = 0$. But if we let $x = i$ and define $i^2 = -1$, we can solve this equation. We call i, an *imaginary number*.

We can now construct the set of *complex numbers* by writing each complex number in the form $a_0 + a_1 i$, where a_0 and a_1 are real numbers and $i^2 = -1$. If two or more variables are next to each other, there is an implied multiplication sign $(\cdot)$ between them, so that $a \cdot b = ab$.

We will have to do some arithmetic with complex numbers and so we need to define operations with these numbers. We start with addition.

Let $m = a_1 + a_2 i$ and $n = b_1 + b_2 i$ be complex numbers. Then we define

$$m + n = (a_1 + b_1) + (a_2 + b_2)i.$$

Multiplication is defined by

$$mn = (a_1 b_1 - a_2 b_2) + (a_1 b_2 + a_2 b_1)i,$$

and scalar multiplication by a real number j is defined by

$$jm = ja_1 + ja_2 i.$$

We also need to find additive and multiplicative identities for our complex numbers. The additive identity will be the complex number

$$0 + 0i$$

and the multiplicative identity will be the complex number

$$1 + 0i.$$

You can use the definitions of addition and multiplication given above to check if these numbers have the required properties.

Complex numbers satisfy the following properties -

Let m, n and p be complex numbers, and let j and k be real numbers. Then there are complex numbers r, s and t such that

1.	$m + n = r$	Closure under Addition
2.	$m + n = n + m$	Commutative Law of Addition
3.	$m + (n + p) = (m + n) + p$	Associative Law of Addition
4.	$m + (0 + 0i) = m$	Additive Identity Property
5.	There's a unique complex number $-m$ such that $m + (-m) = 0 + 0i$	Additive Inverse Property
6.	If $m + n = m + p$, then $n = p$	Cancellation Law of Addition
7.	$jm = s$	Closure under Scalar Multiplication
8.	$jm = mj$	Commutative Law of Scalar Multiplication
9.	$j(km) = (jk)m$	Associative Law of Scalar Multiplication
10.	$(j + k)m = jm + km$	Distributive Property
11.	$j(m + n) = jm + jn$	
12.	$mn = t$	Closure under Multiplication
13.	$mn = nm$	Commutative Law of Multiplication
14.	$m(np) = (mn)p$	Associative Law of Multiplication
15	$m(1 + 0i) = m$	Identity Element for Multiplication
16.	$(m + n)p = mp + np$	Distributive Law
17.	$m - n = p$	Closure under subtraction

We have not been able to define a sensible order for complex numbers.

Theorem 1.7.1:

Two complex numbers are equal, $a_1 + a_2 i = b_1 + b_2 i$, if and only if $a_1 = b_1$ and $a_2 = b_2$. Here a_1, a_2, b_1, and b_2 are real numbers and $i^2 = -1$.

Proof:

Let a_1, a_2, b_1, and b_2 be real numbers and suppose that $i^2 = -1$.

Suppose that $a_1 + a_2 i = b_1 + b_2 i$. Then

$$a_1 - b_1 = b_2 i - a_2 i$$
$$= (b_2 - a_2) i.$$

Squaring both sides yields

$$(a_1 - b_1)^2 = ((b_2 - a_2) i)^2$$
$$= (b_2 - a_2)^2 i^2$$
$$= -(b_2 - a_2)^2$$

Since $i^2 = -1$. Rearranging gives

$$(a_1 - b_1)^2 + (b_2 - a_2)^2 = 0.$$

Since a_1, a_2, b_1, and b_2 are real numbers, this can only happen if $a_1 = b_1$ and $a_2 = b_2$.

The first part of the "if and only if statement" has been proved , now we need to prove the second part.

Suppose $a_1 = b_1$ and $a_2 = b_2$. Then multiplying by i gives

$$a_2 i = b_2 i$$

Thus, after adding $a_1 = b_1$ to both sides, becomes

$$a_1 + a_2 i = b_1 + b_2 i,$$

as is required.

Therefore, $a_1 + a_2 i = b_1 + b_2 i$, if and only if $a_1 = b_1$ and $a_2 = b_2$.

End of proof.

Because we can say that $a_1 + a_2 i = b_1 + b_2 i$ if and only if $a_1 = b_1$ and $a_2 = b_2$, we say that the numbers 1 and i are *linearly independent*. Thus they will form a *basis set* for a *two-dimensional number system*. We can then use the *ordered pair* (a_1, a_2) in the real 2-dimensional plane to represent the complex number $a_1 + a_2 i$.

We lose partial orderings of numbers in 2-dimensional and higher dimensional number systems.

1.8 Another 2-Dimensional Number System

The complex number system is one example of a 2-dimensional number system, but there is another 2-dimensional number system, the *Cartesian coordinate system*. This 2-dimensional number system will be defined by the set of all ordered pairs (x, y), such that $(a_1, a_2) = (b_1, b_2)$ if and only if $a_1 = b_1$ and $a_2 = b_2$, where a_1, a_2, b_1, and b_2 are real numbers. So, given any two numbers in this new 2-dimensional number system, $m = (a_1, a_2)$, and $n = (b_1, b_2)$, we can define arithmetic operations on them which are similar to the operations that have been defined for the complex numbers.

Just like in the complex number system, we'll define

$$m + n = (a_1 + b_1, a_2 + b_2)$$

and we'll also define scalar multiplication by a real number j by

$$jm = (ja_1, ja_2)$$

Unlike in the complex number system, we will not define multiplication of our 2-dimensional numbers. Instead, we will define an operation called the dot product which is defined by -

$$m \cdot n = a_1 b_1 + a_2 b_2$$

The distance between the two numbers will be defined as

$$d = \sqrt{(a_1 - b_1)^2 + (a_2 - b_2)^2}$$

My new 2-dimensional numbers will have the following properties:

Let m, n and p be any of my new 2-dimensional numbers, and let j and k be real numbers. Then there are 2-dimensional numbers r and s such that:

1.	$m + n = r$	Closure under Addition
2.	$m + n = n + m$	Commutative Law of Addition
3.	$m + (n + p) = (m + n) + p$	Associative Law of Addition
4.	$m + (0,0) = m$	Additive Identity Property
5.	There is a unique number $-m$ such that $m + (-m) = (0,0)$	Additive Inverse Property
6.	If $m + n = m + p$, then $n = p$	Cancellation Law of Addition
7.	$jm = s$	Closure under Scalar Multiplication
8.	$jm = mj$	Commutative Law of Scalar Multiplication
9.	$j(km) = (jk)m$	Associative Law of Scalar Multiplication
10.	$(j + k)m = jm + km$	Distributive Property
11.	$j(m + n) = jm + jn$	Distributive Property
12.	$m \cdot n = n \cdot m$	Commutative Law of the Dot Product
13.	$m \cdot (n + p) = m \cdot n + m \cdot p$	Distributive Property of the Dot Product
14.	$j(m \cdot n) = (jm) \cdot n = m \cdot (jn)$	

1.9 A 3-Dimensional Number System -
Modeled After Our Own 3-Dimensional Physical World

Creating a *3-dimensional number system* is like creating a 2-dimensional number system. The elements of 3-dimensional number systems will be numbers of the form $m = (x, y, z)$, where the variables x, y, and z are real numbers representing positions in our own 3-dimensional space. We can say that $(a_1, a_2, a_3) = (b_1, b_2, b_3)$ if and only if $a_1 = b_1$, $a_2 = b_2$, and $a_3 = b_3$. We can then show that the dimensions are linearly independent of one another. I am not forming a system of 3-dimensional vectors, but of 3-dimensional numbers. Remember the difference between a number and a vector is that a number only has a size, whereas a vector has both a magnitude and a direction. The number m is not the same as the vector $\overline{m}$. Taking any two 3-dimensional numbers $m = (a_1, a_2, a_3)$ and $n = (b_1, b_2, b_3)$, where a_1, a_2, a_3, b_1, b_2, and b_3 are real numbers, and there is any real number, j then we can define addition and scalar multiplication similar to the way we had defined for our 2-dimensional number system in the earlier section.

So we will define

$$m + n = (a_1 + b_1, a_2 + b_2, a_3 + b_3)$$

and

$$jm = (ja_1, ja_2, ja_3).$$

Next we shall define the dot product by

$$m \cdot n = a_1 b_1 + a_2 b_2 + a_3 b_3.$$

A new operation, called the cross product, will be defined by

$$m \times n = (a_2 b_3 - a_3 b_2, a_3 b_1 - a_1 b_3, a_1 b_2 - a_2 b_1)$$

and the distance between the two numbers m and n will be defined by

$$d = \sqrt{(a_1 - b_1)^2 + (a_2 - b_2)^2 + (a_3 - b_3)^2}.$$

The 3-dimensional number system that I propose will have the following properties:

Let m, n and p be 3-dimensional numbers, and let j and k be any constant real numbers. Then there are 3-dimensional real numbers r, s and t such that

1.	$m + n = r$	Closure under Addition
2.	$m + n = n + m$	Commutative Law of Addition
3.	$m + (n + p) = (m + n) + p$	Associative Law of Addition
4.	$m + (0,0,0) = m$	Additive Identity Property
5.	There is a unique number $-m$ such that $m + (-m) = (0,0,0)$	Additive Inverse Property
6.	If $m + n = m + p$, then $n = p$	Cancellation Law of Addition
7.	$jm = s$	Closure under Scalar Multiplication
8.	$jm = mj$	Commutative Law of Scalar Multiplication
9.	$j(km) = (jk)m$	Associative Law of Scalar Multiplication
10.	$(j + k)m = jm + km$	Distributive Property
11.	$j(m + n) = jm + jn$	Distributive Property
12.	$m \cdot n = n \cdot m$	Commutative Law of the Dot Product
13.	$m \cdot (n + p) = m \cdot n + m \cdot p$	Distributive Property of the Dot Product.
14.	$m \times n = t$	Closure under Cross Product
15.	$m \times n = -(n \times m)$	
16.	$m \times (n + p) = (m \times n) + (m \times p)$	
17.	$j(m \times n) = (jm) \times n = m \times (jn)$	
18.	$m \times m = (0,0,0)$	
19.	$m \cdot (n \times p) = (m \times n) \cdot p$	

20

1.10 Another 3-Dimensional Number System

The preceding 3-dimensional number system, based on our 3-dimensional physical world, has a drawback that it cannot be turned into a 4-dimensional number system easily because of the cross product. The question of creating another 3-dimensional number system that can easily be turned into a 4-dimensional number system is similar to that of creating a 2-dimensional number system that can easily be turned into a 3-dimensional number system.

The elements of my new 3-dimensional number systems will be 3-dimensional numbers, $m = (x, y, z)$, where the variables x, y, and z are real numbers that are not linearly related to one another.

We shall say that $(a_1, a_2, a_3) = (b_1, b_2, b_3)$ if and only if the real numbers a_1, a_2, a_3, b_1, b_2, and b_3, satisfy the conditions that $a_1 = b_1$, $a_2 = b_2$, and $a_3 = b_3$. We can then see that the dimensions are linearly independent of one another. Given any two numbers, say $m = (a_1, a_2, a_3)$ and $n = (b_1, b_2, b_3)$, we'll again define addition by

$$m + n = (a_1 + b_1, a_2 + b_2, a_3 + b_3).$$

Scalar multiplication by a real number j will be defined by

$$jm = (ja_1, ja_2, ja_3)$$

and we'll define the dot product by

$$m \cdot n = a_1 b_1 + a_2 b_2 + a_3 b_3$$

Instead of the cross product we'll define an operation called the star product by

$$m * n = (a_1 b_1, a_2 b_2, a_3 b_3)$$

The distance between these two numbers will be defined by

$$d = \sqrt{(a_1 - b_1)^2 + (a_2 - b_2)^2 + (a_3 - b_3)^2}$$

The 3-dimensional number system that I propose will also have the following properties:

Let m, n and p be any number in my 3-dimensional numbers, and let j and k be any real numbers. Then there are real numbers r, s and t such that

1.	$m + n = r$	Closure under Addition
2.	$m + n = n + m$	Commutative Law of Addition
3.	$m + (n + p) = (m + n) + p$	Associative Law of Addition
4.	$m + (0,0,0) = m$	Additive Identity Property
5.	There is a unique number –m such that $m + (-m) = (0,0,0)$	Additive Inverse Property
6.	If $m + n = m + p$, then $n = p$	Cancellation Law of Addition
7.	$jm = s$	Closure under Scalar Multiplication
8.	$jm = mj$	Commutative Law of Scalar Multiplication
9.	$j(km) = (jk)m$	Associative Law of Scalar Multiplication
10.	$(j + k)m = jm + km$	Distributive Property
11.	$j(m + n) = jm + jn$	Distributive Property
12.	$m \cdot n = n \cdot m$	Commutative Law of the Dot Product
13.	$m \cdot (n + p) = m \cdot n + m \cdot p$	Distributive Property of the Dot Product
14.	$m * n = t$	Closure under Star Product
15.	$m * n = n * m$	Commutative Law for the Star Product
16.	$(m * n) * p = m * (n * p)$	Associative Law for the Star Product
17.	$(m + n) * p = (m * p) + (n * p)$	Distributive Property for the Star Product

1.11 A 4-Dimensional Number System

Creating a 4-dimensional number system is like creating a 3-dimensional number system. The elements of my 4-dimensional number systems will be numbers of the form $m = (w, x, y, z)$, where the variables w, x, y, and z are real numbers. We will say that $(a_1, a_2, a_3, a_4) = (b_1, b_2, b_3, b_4)$, if and only if $a_1 = b_1$, $a_2 = b_2$, $a_3 = b_3$ and $a_4 = b_4$. Given any two numbers, say $m = (a_1, a_2, a_3, a_4)$ and $n = (b_1, b_2, b_3, b_4)$, we'll define addition by

$$m + n = (a_1 + b_1, a_2 + b_2, a_3 + b_3, a_4 + b_4)$$

Scalar multiplication by the real number j will be defined by

$$jm = (ja_1, ja_2, ja_3, ja_4)$$

and we'll define the dot product by

$$m \cdot n = a_1 b_1 + a_2 b_2 + a_3 b_3 + a_4 b_4$$

The star product will be defined by

$$m * n = (a_1 b_1, a_2 b_2, a_3 b_3, a_4 b_4),$$

and the distance between m and n will be defined as

$$d = \sqrt{(a_1 - b_1)^2 + (a_2 - b_2)^2 + (a_3 - b_3)^2 + (a_4 - b_4)^2}$$

My 4-dimensional number system will also have the following properties:

Let m, n and p be any number in my 4-dimensional numbers, and let j and k be any real numbers. Then there are 4-dimensional numbers r, s and t such that:

1.	$m+n=r$	Closure under Addition
2.	$m+n=n+m$	Commutative Law of Addition
3.	$m+(n+p)=(m+n)+p$	Associative Law of Addition
4.	$m+(0,0,0,0)=m$	Additive Identity Property
5.	There's a unique number $-m$ such that $m+(-m)=(0,0,0,0)$	Additive Inverse Property
6.	If $m+n=m+p$, then $n=p$	Cancellation Law of Addition
7.	$jm=s$	Closure under Scalar Multiplication
8.	$jm=mj$	Commutative Law of Scalar Multiplication
9.	$j(km)=(jk)m$	Associative Law of Scalar Multiplication
10.	$(j+k)m=jm+km$	Distributive Property
11.	$j(m+n)=jm+jn$	Distributive Property
12.	$m \cdot n=n \cdot m$	Commutative Law of the Dot Product
13.	$m \cdot (n+p)=m \cdot n+m \cdot p$	Distributive Property of the Dot Product
14.	$j(m \cdot n)=(jm) \cdot n=m \cdot (jn)$	
15.	$m*n=t$	Closure under Star Product
16.	$m*n=n*m$	Commutative Law for the Star Product
17.	$(m*n)*p=m*(n*p)$	Associative Law for the Star Product
18.	$(m+n)*p=(m*p)+(n*p)$	Distributive Property for the Star Product

Dream! Make your own number systems, with your own operations, and properties, all with nothing.

Part II

The Imprecise Numbers

Chapter 2

Imprecise Numbers as Collections of Precise Numbers

2.1 Basics of Imprecise Numbers

In this chapter, we will be focusing on how to use precise numbers to describe *imprecise numbers*. Remember that an imprecise number is a number that is close in value to the precise number and has the same numeral symbol. There may be more than one precise number that describes a given imprecise number and there may be some precise numbers that can describe it better than others.

2.2 Definition of Collection

In a set of numbers, say, $A = \{a_1, a_2, a_3, \ldots, a_n\}$, no two elements are the same. A *collection* will be defined as grouping of numbers in which any number can be repeated any number of times. We will denote collections using these brackets, $[\![\ ]\!]$. A collection X that contains all the numbers, $x_1, x_2, x_3, \ldots, x_m$, any of which may be repeated any number of times, will be written as $X = [\![x_1, x_2, x_3, \ldots, x_m]\!]$. Each entry in the collection has an equal probability of being an acceptable value for X. So, the more often a given number is repeated, the better its chance of describing the imprecise number. The probability that each entry x_i in the collection, $X = [\![x_1, x_2, x_3, \ldots, x_m]\!]$, ignoring repetitions, will best describe the imprecise number will be $\dfrac{1}{m}$, and, if an entry is repeated k times, the probability that its value best describes the imprecise number will be $\dfrac{k}{m}$.

Example 2.2.1

A collection of precise numbers describing the imprecise number A is $A = [\![0, 1, 1, 2]\!]$. Here, the probability that 0 is the best description of A is 0.25, the probability that 1 is the best value for A is 0.5, and the probability that 2 is the best value for A is 0.25.

Because an imprecise number might be best described by a large number of precise numbers, possibly even an infinitely large number, it would be helpful to be able to generalize all the imprecise numbers by using either one or two precise numbers. Any generalization will cause the imprecise number to be even more imprecise. It would thus be best to work with the collection of all precise numbers that truly describes the value, but in most cases it is impossible to use the collection of all precise numbers. In the study of Statistics, which describes large collections of numbers, it has been found that there are two numbers that works well to describe large collections of data. These two numbers are the mean μ_X and the standard deviation, σ_X.

2.3 The Mean of an Imprecise Number

The *mean* of a finite collection X is the sum of the individual values in the collection (including repeated values), divided by the number of elements in that collection. It is denoted by, μ_X, so that.

$$\mu_X = \frac{\sum_{i=1}^{m} x_i}{m}$$

$$= \frac{x_1 + x_2 + \cdots + x_m}{m},$$

Where, m, is the total number of elements in the collection, X.

Example 2.3.1

Suppose that $A = [\![0, 1, 1, 2]\!]$. Find m, μ_A.

Solution:

A contains 4 values, and so $m = 4$. We calculate

$$\mu_A = \frac{\sum_{i=1}^{4} a_i}{4}$$

$$= \frac{0 + 1 + 1 + 2}{4}$$

$$= \frac{4}{4}$$

$$= 1.$$

2.4 The Standard Deviation of an Imprecise Number

Let $E_i = x_i - \mu_X$. You can think of E_i as the error caused by replacing the ith element of the imprecise number (considered to be a precise number) by the value μ_X. You then square this error to give E_i^2, and take the mean of the squared errors E_i^2, which would be $\dfrac{\sum_{i=1}^{m} E_i^2}{m}$.

We define the standard deviation of the imprecise number, X, denoted by σ_X, to be the square root of $\dfrac{\sum_{i=1}^{m} E_i^2}{m}$, that is,

$$\sigma_X = \sqrt{\frac{\sum_{i=1}^{m} E_i^2}{m}}$$

$$= \sqrt{\frac{\sum_{i=1}^{m} \left(x_i - \mu_A\right)^2}{m}}$$

$$= \sqrt{\frac{\sum_{i=1}^{m} \left(x_i - \dfrac{\sum_{i=1}^{m} x_i}{m}\right)^2}{m}}.$$

Example 2.4.1

Suppose that $A = [\![\, 0, 1, 1, 2 \,]\!]$. Find m, σ_A.

Solution:
In Example 2.3.1 we calculated that $m = 4$. So

$$\sigma_A = \sqrt{\frac{\sum_{i=1}^{m} \left(a_i - \mu_A\right)^2}{m}}$$

$$= \sqrt{\frac{\left(0-1\right)^2 + \left(1-1\right)^2 + \left(1-1\right)^2 + \left(2-1\right)^2}{4}}$$

$$= \frac{\sqrt{2}}{2}.$$

2.5 Addition of a Real Number to a Collection

When a real number is added to a collection, the result then becomes a collection in which that real number has been added to each of the elements of the original collection.

If $X = [\![\, x_1, x_2, x_3, \ldots, x_m \,]\!]$ and k is a real number, then $X + k = [\![\, x_1 + k, x_2 + k, \cdots, x_m + k \,]\!]$.

Example 2.5.1

Suppose $B = A + 1$ where $A = [\![\, 0, 1, 1, 2 \,]\!]$. Find B.

Solution:

$$B = A + 1$$
$$= [\![\, 0, 1, 1, 2 \,]\!] + 1$$
$$= [\![\, 0+1, 1+1, 1+1, 2+1 \,]\!]$$
$$= [\![\, 1, 2, 2, 3 \,]\!].$$

2.6 The Mean of a Collection after the Addition of a Real Number

Theorem 2.6.1:

Suppose the mean of the collection $X = [\![\, x_1, x_2, x_3, \ldots, x_m \,]\!]$ is μ_X, and that k is a real number. Then the mean of the collection $X + k = [\![\, x_1 + k, x_2 + k, \ldots, x_m + k \,]\!]$ is given by $\mu_{X+k} = \mu_X + k$.

Proof:

Let $A = [\![\, a_1, a_2, \ldots, a_m \,]\!]$ be a collection with mean μ_A and let k be a real number. Then

$$\sum_{i=1}^{m} (a_i + k) = (a_1 + k + a_2 + k + \cdots + a_m + k)$$
$$= a_1 + a_2 + \cdots + a_m + mk$$
$$= \left(\sum_{i=1}^{m} a_i \right) + mk$$
$$= m\mu_A + mk$$
$$= m(\mu_A + k).$$

$$\mu_{A+k} = \frac{\sum_{i=1}^{m} (a_i + k)}{m}$$
$$= \frac{m(\mu_A + k)}{m}$$
$$= \mu_A + k,$$

as required.

End of proof.

When adding a real number to a collection, the mean of the collection is shifted by the real number k to give $\mu_{X+k} = \mu_X + k$.

Example 2.6.1

Suppose that $B = A + 1$, where $A = [\![0, 1, 1, 2]\!]$. Find B, m, μ_A, μ_B

Solution:

In Example 2.5.1, we saw that $B = [\![1, 2, 2, 3]\!]$. It is easy to see that $m = 4$.
In Example 2.3, we calculated the mean, $\mu_A = 1$. By Theorem 2.6.1,

$$\begin{aligned}
\mu_B &= \mu_A + 1 \\
&= 1 + 1 \\
&= 2.
\end{aligned}$$

Alternatively, we can calculate μ_B using the definition of the mean as follows:

$$\begin{aligned}
\mu_B &= \frac{\displaystyle\sum_{i=1}^{4} b_i}{m} \\
&= \frac{1 + 2 + 2 + 3}{4} \\
&= \frac{8}{4} \\
&= 2.
\end{aligned}$$

2.7 The Standard Deviation after the Addition of a Real Number

Theorem 2.7.1:

Suppose that the standard deviation of a collection $X = [\![x_1, x_2, x_3, \ldots, x_m]\!]$ is σ_X, and that k is a real number. Then the standard deviation of the collection $X + k = [\![x_1 + k, x_2 + k, \ldots, x_m + k]\!]$ is $\sigma_{X+k} = \sigma_X$.

Proof:

Suppose that σ_A is the standard deviation of, $A = [\![a_1, a_2, \ldots, a_m]\!]$, and k is a real number. Then, by definition of the standard deviation, we have

$$\begin{aligned}
\sigma_{A+k} &= \sqrt{\frac{\displaystyle\sum_{i=1}^{m}\left((a_i + k) - \mu_{A+k}\right)^2}{m}} \\
&= \sqrt{\frac{\displaystyle\sum_{i=1}^{m}\left((a_i + k) - (\mu_A + k)\right)^2}{m}}.
\end{aligned}$$

by Theorem 2.6.1.

Applying the distributive law,

$$\sigma_{A+k} = \sqrt{\frac{\sum_{i=1}^{m}\left(a_i + k - \mu_A - k\right)^2}{m}}$$

$$= \sqrt{\frac{\sum_{i=1}^{m}\left(a_i - \mu_A\right)^2}{m}}$$

$$= \sigma_A \quad,$$

as required.

End of proof.

Adding a real number to a collection does not change the standard deviation.

Example 2.7.1

Suppose $B = A + 1$, where $A = [\![0, 1, 1, 2]\!]$. Find B, m, μ_A, μ_B, σ_A, σ_B.

Solution:

Now, from Example 2.5.1 we have $B = [\![1, 2, 2, 3]\!]$ and $m = 4$.

By Examples 2.31 and 2.5.1, $\mu_A = 1$ and $\mu_B = 2$.

In Example 2.4.1, we calculated that, $\sigma_A = \dfrac{\sqrt{2}}{2}$. So, by Theorem 2.7.1,

$$\sigma_B = \sigma_{A+1}$$

$$= \sigma_A$$

$$= \frac{\sqrt{2}}{2}.$$

Alternatively, we can use the definition of σ_B to verify that

$$\sigma_B = \sqrt{\frac{\left(1-2\right)^2 + \left(2-2\right)^2 + \left(2-2\right)^2 + \left(3-2\right)^2}{4}}$$

$$= \frac{\sqrt{2}}{2}.$$

2.8 Subtraction of a Real Number from a Collection

When we subtract a real number from a collection, the result is a collection in which each element of the original collection has that real number subtracted from it.

If $X = [\![x_1, x_2, x_3, \ldots, x_m]\!]$ and k is a real number, then $X - k = [\![x_1 - k, x_2 - k, \cdots, x_m - k]\!]$.

Example 2.8.1

Suppose that $A = [\![0, 1, 1, 2]\!]$. Find $C = A - 1$.

Solution:

$$C = A - 1$$
$$= [\![0, 1, 1, 2]\!] - 1$$
$$= [\![0 - 1, 1 - 1, 1 - 1, 2 - 1]\!]$$
$$= [\![-1, 0, 0, 1]\!].$$

2.9 The Mean and Standard Deviation of a Collection after Subtraction of a Real Number

Because addition is defined for all the collections X and all real numbers, k, we can think of subtraction as addition with the additive inverse of k, i.e., as $X - k = X + (-k)$. Then, applying the results of Theorems 2.6.1 and 2.7.1, we see that the mean of $X - k$ is

$$\mu_{X-k} = \mu_X - k$$

and the standard deviation of $X - k$ is

$$\sigma_{X-k} = \sigma_X.$$

Example 2.9.1

Suppose that $A = [\![0, 1, 1, 2]\!]$ and $C = A - 1$. Find m, μ_C, σ_C

Solution:
Now, by Example 2.8.1 we have $C = [\![-1, 0, 0, 1]\!]$ and $m = 4$.
By Theorem 2.6.1 and Example 2.4.1, we have

$$\mu_C = \mu_A - 1$$
$$= 1 - 1$$
$$= 0.$$

Alternatively, we may use the definition of μ to verify that

$$\mu_C = \frac{\sum_{i=1}^{4} c_i}{4}$$

$$= \frac{-1+0+0+1}{4}.$$

$$= \frac{0}{4}$$

$$= 0.$$

Further, by Theorem 2.7.1 and Example 2.4.1, we calculate

$$\sigma_C = \sigma_{A-1}$$

$$= \sigma_A .$$

$$= \frac{\sqrt{2}}{2}$$

Alternatively, we may use the definition of σ to verify that

$$\sigma_C = \sqrt{\frac{(-1-0)^2 + (0-0)^2 + (0-0)^2 + (1-0)^2}{4}}$$

$$= \frac{\sqrt{2}}{2}.$$

2.10 Definition of the Addition of Two Collections

When we add two collections, each element of one collection is added to each element of the other collection because these sums all have equal likelihoods of being acceptable values for the new collection. Thus, if the first collection had m elements and the second collection had n elements, the sum of the two collections would have mn elements.

If $X = [\![x_1, x_2, x_3, \ldots, x_m]\!]$ and $Y = [\![y_1, y_2, y_3, \ldots, y_n]\!]$, then we define

$$X + Y = \left[\!\!\left[\begin{array}{l} x_1 + y_1, x_1 + y_2, x_1 + y_3, \cdots, x_1 + y_n, \\ x_2 + y_1, x_2 + y_2, x_2 + y_3, \cdots, x_2 + y_n, \\ x_3 + y_1, x_3 + y_2, x_3 + y_3, \cdots, x_3 + y_n, \\ \vdots \qquad \vdots \qquad \vdots \qquad \ddots \qquad \vdots \\ x_m + y_1, x_m + y_2, x_m + y_3, \cdots, x_m + y_n \end{array} \right]\!\!\right].$$

Example 2.10.1

Suppose $A = [\![0, 1, 1, 2]\!]$ and $C = [\![-1, 0, 0, 1]\!]$. Find $F = A + C$.

Solution:

By definition,

$$F = A + C$$
$$= [\![\ a_1 + c_1, a_1 + c_2, a_1 + c_3, a_1 + c_4,$$
$$a_2 + c_1, a_2 + c_2, a_2 + c_3, a_2 + c_4,$$
$$a_3 + c_1, a_3 + c_2, a_3 + c_3, a_3 + c_4,$$
$$a_4 + c_1, a_4 + c_2, a_4 + c_3, a_4 + c_4\ ]\!]$$

$$= [\![\ 0 + (-1), 0 + 0, 0 + 0, 0 + 1,$$
$$1 + (-1), 1 + 0, 1 + 0, 1 + 1,$$
$$1 + (-1), 1 + 0, 1 + 0, 1 + 1,$$
$$2 + (-1), 2 + 0, 2 + 0, 2 + 1\ ]\!]$$
$$= [\![-1, 0, 0, 1, 0, 1, 1, 2, 0, 1, 1, 2, 1, 2, 2, 3]\!].$$

2.11 The Mean of the Addition of Two Collections

Theorem 2.11.1: Suppose that the mean of collection $X = [\![x_1, x_2, x_3, \ldots, x_m]\!]$ is μ_X and the mean of collection $Y = [\![y_1, y_2, y_3, \ldots, y_n]\!]$ is μ_Y. Then the mean of the collection X+Y is
$$\mu_{X+Y} = \mu_X + \mu_Y.$$

Proof:

Suppose $X = [\![x_1, x_2, x_3, \ldots, x_m]\!]$ and $Y = [\![y_1, y_2, y_3, \ldots, y_n]\!]$. Then

$$\mu_X = \frac{\sum_{i=1}^{m} x_i}{m}, \ \mu_Y = \frac{\sum_{i=1}^{n} y_i}{n}, \text{ and}$$

$$X + Y = \begin{bmatrix} x_1 + y_1, x_1 + y_2, x_1 + y_3, \cdots, x_1 + y_n, \\ x_2 + y_1, x_2 + y_2, x_2 + y_3, \cdots, x_2 + y_n, \\ x_3 + y_1, x_3 + y_2, x_3 + y_3, \cdots, x_3 + y_n, \\ \vdots \quad \vdots \quad \vdots \quad \ddots \quad \vdots \\ x_m + y_1, x_m + y_2, x_m + y_3, \cdots, x_m + y_n \end{bmatrix} \text{ then}$$

$$\mu_{X+Y} = \frac{\displaystyle\sum_{i=1}^{m}\sum_{j=1}^{n}(x_i + y_j)}{mn}$$

$$\mu_{X+Y} = \frac{n\displaystyle\sum_{i=1}^{m} x_i + m\sum_{j=1}^{n} y_j}{mn}$$

$$\mu_{X+Y} = \frac{n\displaystyle\sum_{i=1}^{m} x_i}{mn} + \frac{m\displaystyle\sum_{j=1}^{n} y_j}{mn}$$

$$\mu_{X+Y} = \frac{\displaystyle\sum_{i=1}^{m} x_i}{m} + \frac{\displaystyle\sum_{j=1}^{n} y_j}{n}$$

Therefore,

$$\mu_{X+Y} = \mu_X + \mu_Y$$

as required.

End of proof.

When two collections are added together, the mean is the sum of the means of the two individual collections, that is $\mu_{X+Y} = \mu_X + \mu_Y$.

Example 2.11.1

Suppose $A = [\![0, 1, 1, 2]\!]$, $C = [\![-1, 0, 0, 1]\!]$ and $F = A + C$. Find μ_F.

Solution:

Now, from Example 2.10.1, we have

$$F = A + C$$
$$= [\![-1, 0, 0, 1, 0, 1, 1, 2, 0, 1, 1, 2, 1, 2, 2, 3]\!].$$

Also, in Examples 2.3.1 and 2.9.1, we found that $\mu_A = 1$ and $\mu_C = 0$. Therefore, by Theorem 2.11.1, we have

$$\mu_F = \mu_A + \mu_C$$
$$= 1 + 0$$
$$= 1.$$

We may also use the definition of μ to verify that

$$\mu_F = \frac{\sum\limits_{i=1}^{m} f_i}{m}$$

$$= \frac{16}{16}$$

$$= 1.$$

2.12 The Standard Deviation of the Sum of Two Collections

Theorem 2.12.1:

Suppose, that the standard deviation of the collection, $X = [\![x_1, x_2, x_3, \ldots, x_m]\!]$ is σ_X and the standard deviation of the collection, $Y = [\![y_1, y_2, y_3, \ldots, y_n]\!]$ is σ_Y. Then the standard deviation of the collection $X + Y$, is $\sigma_{X+Y} = \sqrt{\sigma_X^2 + \sigma_Y^2}$.

Proof:

Let $X = [\![x_1, x_2, x_3, \ldots, x_m]\!]$, $Y = [\![y_1, y_2, y_3, \ldots, y_n]\!]$ and write

$$X + Y = \begin{Vmatrix} x_1 + y_1, x_1 + y_2, x_1 + y_3, \cdots, x_1 + y_n, \\ x_2 + y_1, x_2 + y_2, x_2 + y_3, \cdots, x_2 + y_n, \\ x_3 + y_1, x_3 + y_2, x_3 + y_3, \cdots, x_3 + y_n, \\ \vdots \quad \vdots \quad \vdots \quad \ddots \quad \vdots \\ x_m + y_1, x_m + y_2, x_m + y_3, \cdots, x_m + y_n \end{Vmatrix}. \quad \text{Then}$$

$$\sigma_{X+Y} = \sqrt{\frac{\sum\limits_{i=1}^{m}\sum\limits_{j=1}^{n}\left((x_i + y_j) - \mu_{X+Y}\right)^2}{mn}}$$

$$\sigma_{X+Y} = \sqrt{\frac{\sum\limits_{i=1}^{m}\sum\limits_{j=1}^{n}\left((x_i + y_j) - (\mu_X + \mu_Y)\right)^2}{mn}}$$

$$\sigma_{X+Y}^2 = \frac{\sum\limits_{i=1}^{m}\sum\limits_{j=1}^{n}\left((x_i + y_j) - (\mu_X + \mu_Y)\right)^2}{mn}$$

We calculate -

$$mn\sigma_{X+Y}^2 = \sum\limits_{i=1}^{m}\sum\limits_{j=1}^{n}\left((x_i + y_j) - (\mu_X + \mu_Y)\right)^2$$

$$mn\sigma_{X+Y}^2 = \sum\limits_{i=1}^{m}\sum\limits_{j=1}^{n}\left((x_i + y_j)^2 - 2(x_i + y_j)(\mu_X + \mu_Y) + (\mu_X + \mu_Y)^2\right)$$

$$mn\sigma^2_{X+Y} = \sum_{i=1}^{m}\sum_{j=1}^{n}\left(\begin{array}{c} \left(x_i^2 + 2x_i y_j + y_j^2\right) \\ -2\left(x_i\mu_X + y_j\mu_X + x_i\mu_Y + y_j\mu_Y\right) \\ +\left(\mu_X^2 + 2\mu_X\mu_Y + \mu_Y^2\right) \end{array} \right)$$

$$mn\sigma^2_{X+Y} = \sum_{i=1}^{m}\sum_{j=1}^{n}\left(\begin{array}{c} \left(x_i^2 - 2x_i\mu_X + \mu_X^2\right) + 2x_i y_j \\ -2\left(y_j\mu_X + x_i\mu_Y\right) \\ +\left(y_j^2 - 2y_j\mu_Y + \mu_Y^2\right) + 2\mu_X\mu_Y \end{array} \right)$$

$$mn\sigma^2_{X+Y} = \sum_{i=1}^{m}\sum_{j=1}^{n}\left(\begin{array}{c} \left(x_i^2 - 2x_i\mu_X + \mu_X^2\right) \\ +2\left(x_i y_j - y_j\mu_X - x_i\mu_Y + \mu_X\mu_Y\right) \\ +\left(y_j^2 - 2y_j\mu_Y + \mu_Y^2\right) \end{array} \right)$$

$$mn\sigma^2_{X+Y} = \sum_{i=1}^{m}\sum_{j=1}^{n}\left(\begin{array}{c} \left(x_i - \mu_X\right)^2 \\ +2\left(x_i - \mu_X\right)\left(y_j - \mu_Y\right) \\ +\left(y_j - \mu_Y\right)^2 \end{array} \right)$$

$$mn\sigma^2_{X+Y} = n\sum_{i=1}^{m}\left(x_i - \mu_X\right)^2 + m\sum_{j=1}^{n}\left(y_j - \mu_Y\right)^2 + 2\left(\sum_{i=1}^{m}\left(x_i - \mu_X\right)\right)\left(\sum_{j=1}^{n}\left(y_j - \mu_Y\right)\right)$$

$$mn\sigma^2_{X+Y} = n\sum_{i=1}^{m}\left(x_i - \mu_X\right)^2 + m\sum_{j=1}^{n}\left(y_j - \mu_Y\right)^2 + 2\left(\left(\sum_{i=1}^{m}(x_i)\right) - m\mu_X\right)\left(\left(\sum_{j=1}^{n}(y_j)\right) - n\mu_Y\right)$$

Dividing the above by mn gives

$$\sigma^2_{X+Y} = \frac{n\sum_{i=1}^{m}\left(x_i - \mu_X\right)^2}{mn} + \frac{m\sum_{j=1}^{n}\left(y_j - \mu_Y\right)^2}{mn} + \frac{2\left(\left(\sum_{i=1}^{m}(x_i)\right) - m\mu_X\right)\left(\left(\sum_{j=1}^{n}(y_j)\right) - n\mu_Y\right)}{mn}$$

$$\sigma^2_{X+Y} = \frac{\sum_{i=1}^{m}\left(x_i - \mu_X\right)^2}{m} + \frac{\sum_{j=1}^{n}\left(y_j - \mu_Y\right)^2}{n} + 2\frac{\left(\left(\sum_{i=1}^{m}(x_i)\right) - m\mu_X\right)}{m} \cdot \frac{\left(\left(\sum_{j=1}^{n}(y_j)\right) - n\mu_Y\right)}{n}$$

$$\sigma^2_{X+Y} = \frac{\sum_{i=1}^{m}\left(x_i - \mu_X\right)^2}{m} + \frac{\sum_{j=1}^{n}\left(y_j - \mu_Y\right)^2}{n} + 2\left(\frac{\sum_{i=1}^{m}(x_i)}{m} - \mu_X\right) \cdot \left(\frac{\sum_{j=1}^{n}(y_j)}{n} - \mu_Y\right)$$

$$\sigma^2_{X+Y} = \frac{\sum_{i=1}^{m}\left(x_i - \mu_X\right)^2}{m} + \frac{\sum_{j=1}^{n}\left(y_j - \mu_Y\right)^2}{n} + 2\left(\mu_X - \mu_X\right)\left(\mu_Y - \mu_Y\right)$$

38

$$\sigma_{X+Y}^2 = \sigma_X^2 + \sigma_Y^2$$

Therefore, $\sigma_{X+Y} = \sqrt{\sigma_X^2 + \sigma_Y^2}$, as required.

End of proof.

The standard deviation of the sum of two collections is the square root of the sum of the squares of the standard deviations of the two individual collections, that is, $\sigma_{X+Y} = \sqrt{\sigma_X^2 + \sigma_Y^2}$.

Example 2.12.1

Suppose $A = [\![0, 1, 1, 2]\!]$, $C = [\![-1, 0, 0, 1]\!]$ and $F = A + C$. Find σ_F.

Solution:

In Example 2.10.1, we showed that $F = A + C = [\![-1, 0, 0, 1, 0, 1, 1, 2, 0, 1, 1, 2, 1, 2, 2, 3]\!]$.

In Examples 2.4.1 and 2.9.1, we saw that $\sigma_A = \dfrac{\sqrt{2}}{2}$ and $\sigma_C = \dfrac{\sqrt{2}}{2}$.

By Theorem 2.11.1, we calculate

$$\begin{aligned}
\sigma_F &= \sigma_{A+C} \\
&= \sqrt{\sigma_A^2 + \sigma_C^2} \\
&= \sqrt{\left(\frac{\sqrt{2}}{2}\right)^2 + \left(\frac{\sqrt{2}}{2}\right)^2} \\
&= \sqrt{\frac{2}{4} + \frac{2}{4}} \\
&= 1.
\end{aligned}$$

Alternatively, we may use the definition of the standard deviation to verify that

$$\begin{aligned}
\sigma_F &= \sqrt{\frac{\sum_{i=1}^{m}(f_i - \mu_F)^2}{m}} \\
&= \sqrt{\frac{\sum(f_i - 1)^2}{16}} \\
&= \sqrt{\frac{(-1-1)^2 + 4(0-1)^2 + 6(1-1)^2 + 4(2-1)^2 + (3-1)^2}{16}} \\
&= \sqrt{\frac{4 + 4 + 0 + 4 + 4}{16}} \\
&= 1
\end{aligned}$$

2.13 Definition of Multiplication of a Collection by a Real Number

When a collection is multiplied by a real number, the result is a collection in which each of the elements of the original collection is multiplied by that real number.

If $X = [\![x_1, x_2, x_3, \ldots, x_m]\!]$ and k is a real number, then $kX = [\![kx_1, kx_2, kx_3, \ldots, kx_m]\!]$.

Example 2.13.1

Suppose $A = [\![0, 1, 1, 2]\!]$, $k = -2$ and $F = kA$. Find F.

Solution:

$$F = kA$$
$$= [\![(-2)(0), (-2)(1), (-2)(1), (-2)(2)]\!]$$
$$= [\![0, -2, -2, -4]\!].$$

2.14 The Mean of a Collection after Multiplication by a Real Number

Theorem 2.14.1: Let the mean of the collection, $X = [\![x_1, x_2, x_3, \ldots, x_m]\!]$ be μ_X, and let k be a real number. Then the mean of the collection, $kX = [\![kx_1, kx_2, kx_3, \ldots, kx_m]\!]$ is $\mu_{kX} = k\mu_X$.

Proof:

Suppose μ_X is the mean of $X = [\![x_1, x_2, x_3, \ldots, x_m]\!]$ and k is a real number. Then

$$\mu_{kX} = \frac{\sum_{i=1}^{m} kx_i}{m}$$

and so,

$$\mu_{kX} = \frac{k\sum_{i=1}^{m} x_i}{m}$$

$$\mu_{kX} = k\frac{\sum_{i=1}^{m} x_i}{m}$$

$$\mu_{kX} = k\mu_X$$

End of proof.

When a collection, X, is multiplied by a real number k, the mean of the new collection is k times the mean of the original collection, that is $\mu_{kX} = k\mu_X$.

Example 2.14.1

Suppose $B = kA$ where $A = [\![0, 1, 1, 2]\!]$ and $k = -2$. Find B, m, μ_A, and μ_B.

Solution:

In Example 2.13.1, we found that $B = [\![0, -2, -2, -4]\!]$ and $m = 4$.
Further, we have $\mu_A = 1$ by Example 2.3.1.
Now, by Theorem 2.14.1,

$$\begin{aligned} \mu_B &= k\mu_A \\ &= (-2)(1) \\ &= -2. \end{aligned}$$

Alternatively, we may use the definition of the mean to verify that,

$$\begin{aligned} \mu_B &= \frac{0 - 2 - 2 - 4}{4} \\ &= -2. \end{aligned}$$

2.15 Standard Deviation after Multiplication by a Real Number

Theorem 2.15.1:

Let the standard deviation of the collection, $X = [\![x_1, x_2, x_3, \ldots, x_m]\!]$ be σ_X, and let k be a real number. Then the standard deviation of the collection, $kX = [\![kx_1, kx_2, kx_3, \ldots, kx_m]\!]$ is $\sigma_{kX} = |k|\sigma_X$.

Proof:

Suppose that σ_X is the standard deviation of $X = [\![x_1, x_2, x_3, \ldots, x_m]\!]$, and k is a real number. Then, by definition of σ_{kX}, we have

$$\sigma_{kX} = \sqrt{\frac{\sum_{i=1}^{m} \left((kx_i) - \mu_{kX} \right)^2}{m}}.$$

Since $\mu_{kX} = k\mu_X$,

$$\sigma_{kX} = \sqrt{\frac{\sum_{i=1}^{m} \left((kx_i) - k\mu_X \right)^2}{m}}$$

$$\sigma_{kX} = \sqrt{\dfrac{\sum\limits_{i=1}^{m} k^2 \left(x_i - \mu_X\right)^2}{m}}$$

$$\sigma_{kX} = \sqrt{\dfrac{k^2 \sum\limits_{i=1}^{m} \left(x_i - \mu_X\right)^2}{m}}$$

$$\sigma_{kX} = |k| \sqrt{\dfrac{\sum\limits_{i=1}^{m} \left(x_i - \mu_X\right)^2}{m}}$$

$$\sigma_{kX} = |k| \sigma_X$$

as required.

End of proof.

When we multiply a collection, X by a real number k, the standard deviation of the new collection is $\sigma_{kX} = |k| \sigma_X$.

Example 2.15.1

Suppose that $B = kA$ where $A = [\![0, 1, 1, 2]\!]$ and $k = -2$. Find B, m, σ_A, and σ_B.

Solution:

In Example 2.13.1, we found that $B = [\![0, -2, -2, -4]\!]$ and $m = 4$.

In Example 2.3.1, we found that $\mu_A = 1$, and in Example 2.14.1, we found that $\mu_B = -2$.

Finally, in Example 2.4.1, we showed that $\sigma_A = \dfrac{\sqrt{2}}{2}$.

Therefore, by Theorem 2.15.1, we have -

$$\begin{aligned}
\sigma_B &= |k| \sigma_A \\
&= |-2| \left(\dfrac{\sqrt{2}}{2}\right) \\
&= \sqrt{2}.
\end{aligned}$$

Alternatively, we may use the definition of the standard deviation to verify that

$$\sigma_B = \sqrt{\frac{\sum \left(b_i - \mu_B\right)^2}{m}}$$

$$= \sqrt{\frac{\left(0-(-2)\right)^2 + \left(-2-(-2)\right)^2 + \left(-2-(-2)\right)^2 + \left(-2-(-4)\right)^2}{4}}$$

$$= \sqrt{\frac{4+0+0+4}{4}}$$

$$= \sqrt{2}.$$

2.16 Definition of the Subtraction of Two Collections

To subtract two collections, we subtract each element of the second collection from each element of the first collection. All of the differences obtained in this way have equal likelihoods of being acceptable values for the new collection. This means that if the first collection has m elements and the second collection has n elements, then the subtraction of the two collections will have mn elements.

If $X = [\![x_1, x_2, x_3, \ldots, x_m]\!]$ and $Y = [\![y_1, y_2, y_3, \ldots, y_n]\!]$, then we define

$$X - Y = \begin{Bmatrix} x_1 - y_1, x_1 - y_2, x_1 - y_3, \cdots, x_1 - y_n, \\ x_2 - y_1, x_2 - y_2, x_2 - y_3, \cdots, x_2 - y_n, \\ x_3 - y_1, x_3 - y_2, x_3 - y_3, \cdots, x_3 - y_n, \\ \vdots \quad \vdots \quad \vdots \quad \ddots \quad \vdots \\ x_m - y_1, x_m - y_2, x_m - y_3, \cdots, x_m - y_n \end{Bmatrix}.$$

Example 2.16.1

Suppose $A = [\![0, 1, 1, 2]\!]$ and $C = [\![-1, 0, 0, 1]\!]$. Find $F = A - C$.

Solution:

$$F = A - C = \begin{Bmatrix} a_1 - c_1, a_1 - c_2, a_1 - c_3, a_1 - c_4, \\ a_2 - c_1, a_2 - c_2, a_2 - c_3, a_2 - c_4, \\ a_3 - c_1, a_3 - c_2, a_3 - c_3, a_3 - c_4, \\ a_4 - c_1, a_4 - c_2, a_4 - c_3, a_4 - c_4 \end{Bmatrix}$$

$$F = A - C = \begin{Bmatrix} 0-(-1), 0-0, 0-0, 0-1, \\ 1-(-1), 1-0, 1-0, 1-1, \\ 1-(-1), 1-0, 1-0, 1-1, \\ 2-(-1), 2-0, 2-0, 2-1 \end{Bmatrix}$$

$$F = [\![1, 0, 0, -1, 2, 1, 1, 0, 2, 1, 1, 0, 3, 2, 2, 1]\!]$$

2.17 The Mean of the Subtraction of Two Collections

Theorem 2.17.1:

Let the mean of collection $X = [\![x_1, x_2, x_3, \ldots, x_m]\!]$, be μ_X and the mean of collection $Y = [\![y_1, y_2, y_3, \ldots, y_n]\!]$, be μ_Y. Then the mean of the collection $X - Y$ is $\mu_{X-Y} = \mu_X - \mu_Y$.

Proof:

Suppose that μ_X is the mean of $X = [\![x_1, x_2, x_3, \ldots, x_m]\!]$ and μ_Y is the mean of $Y = [\![y_1, y_2, y_3, \ldots, y_n]\!]$. Then

$$\mu_{X-Y} = \mu_{X+(-1)Y}$$
$$\mu_{X-Y} = \mu_X + \mu_{(-1)Y} \qquad \text{by Theorem 2.11.1}$$
$$\mu_{X-Y} = \mu_X - \mu_Y \qquad \text{by Theorem 2.14.1,}$$

as required.

End of proof.

Example 2.17.1

Suppose $A = [\![0, 1, 1, 2]\!]$, $C = [\![-1, 0, 0, 1]\!]$ and $F = A - C$. Find μ_F.

Solution:

Now, we found that $\mu_A = 1$ in Example 2.3.1 and $\mu_C = 0$ in Example 2.9.1. Therefore, by Theorem 2.17.1

$$\mu_F = \mu_A - \mu_C$$
$$= 1 - 0 \qquad .$$
$$= 1.$$

2.18 The Standard Deviation of the Subtraction of Two Collections

Theorem 2.18.1:

Let the standard deviation of the collection $X = [\![x_1, x_2, x_3, \ldots, x_m]\!]$, be σ_X and the standard deviation of the collection $Y = [\![y_1, y_2, y_3, \ldots, y_n]\!]$, be σ_Y. Then the standard deviation of $X - Y$, is $\sigma_{X-Y} = \sqrt{\sigma_X^2 + \sigma_Y^2}$.

Proof:

Suppose that $X = [\![x_1, x_2, x_3, \ldots, x_m]\!]$, $Y = [\![y_1, y_2, y_3, \ldots, y_n]\!]$. Then

$$\sigma_{X-Y} = \sigma_{X+(-1)Y}$$
$$\sigma_{X-Y} = \sqrt{\sigma_X^2 + \sigma_{(-1)Y}^2} \qquad \text{by Theorem 2.11.1}$$

$$\sigma_{X-Y} = \sqrt{\sigma_X^2 + \left(|-1|\sigma_Y\right)^2} \qquad \text{by Theorem 2.15.1.}$$

Therefore,

$$\sigma_{X-Y} = \sqrt{\sigma_X^2 + \sigma_Y^2}\,,$$

as required.

End of proof.

Example 2.18.1

Suppose $A = [\![0, 1, 1, 2]\!]$, $C = [\![-1, 0, 0, 1]\!]$ and $F = A - C$. Find σ_F.

Solution:

Now, by Examples 2.4.1 and 2.9.1, $\sigma_A = \dfrac{\sqrt{2}}{2}$ and $\sigma_C = \dfrac{\sqrt{2}}{2}$.

By Theorem 2.18.1, we have

$$\begin{aligned}
\sigma_F &= \sigma_{A-C} \\
&= \sqrt{\sigma_A^2 + \sigma_C^2} \\
&= \sqrt{\left(\frac{\sqrt{2}}{2}\right)^2 + \left(\frac{\sqrt{2}}{2}\right)^2} \\
&= \sqrt{\frac{2}{4} + \frac{2}{4}} \\
&= 1.
\end{aligned}$$

Alternatively, we may calculate

$$F = A - C = \begin{bmatrix} 0-(-1), 0-0, 0-0, 0-1, \\ 1-(-1), 1-0, 1-0, 1-1, \\ 1-(-1), 1-0, 1-0, 1-1, \\ 2-(-1), 2-0, 2-0, 2-1 \end{bmatrix}$$

$$F = [\![1, 0, 0, -1, 2, 1, 1, 0, 2, 1, 1, 0, 3, 2, 2, 1]\!]$$

Then, by Example 2.17.1, we have

$$\begin{aligned}
\mu_F &= \mu_A - \mu_C \\
&= 1 - 0 \\
&= 1.
\end{aligned}$$

By the definition of the standard deviation,

$$\sigma_F = \sqrt{\frac{\sum\limits_{i=1}^{16}(f_i-1)^2}{16}}$$

$$= \sqrt{\frac{(-1-1)^2+4(0-1)^2+6(1-1)^2+4(2-1)^2+(3-1)^2}{16}}$$

$$= \sqrt{\frac{4+4+0+4+4}{16}}$$

$$= 1.$$

2.19 Definition of the Multiplication of Two Collections

When two collections are multiplied, each element of one collection is multiplied once by each element of the other collection. All of these possible products have equal likelihoods of being acceptable values for the new collection. Thus, if the first collection has m elements and the second collection has n elements, then the product of the two collections will have mn elements.

If $X = [\![x_1, x_2, x_3, \ldots, x_m]\!]$ and $Y = [\![y_1, y_2, y_3, \ldots, y_n]\!]$, then

$$XY = \begin{Bmatrix} x_1y_1, x_1y_2, x_1y_3, \cdots, x_1y_n, \\ x_2y_1, x_2y_2, x_2y_3, \cdots, x_2y_n, \\ x_3y_1, x_3y_2, x_3y_3, \cdots, x_3y_n, \\ \vdots \quad \vdots \quad \vdots \quad \ddots \quad \vdots \\ x_my_1, x_my_2, x_my_3, \cdots, x_my_n \end{Bmatrix}.$$

Example 2.19.1

Let $A = [\![0, 1, 1, 2]\!]$ and $C = [\![-1, 0, 0, 1]\!]$.   Find $F = AC$.

Solution:

$$F = AC = \begin{Bmatrix} a_1c_1, a_1c_2, a_1c_3, a_1c_4, a_2c_1, a_2c_2, a_2c_3, a_2c_4, \\ a_3c_1, a_3c_2, a_3c_3, a_3c_4, a_4c_1, a_4c_2, a_4c_3, a_4c_4 \end{Bmatrix}$$

$$F = \begin{Bmatrix} (0)(-1),(0)(0),(0)(0),(0)(1),(1)(-1),(1)(0),(1)(0),(1)(1), \\ (1)(-1),(1)(0),(1)(0),(1)(1),(2)(-1),(2)(0),(2)(0),(2)(1) \end{Bmatrix}$$

$$F = [\![0,0,0,0,-1,0,0,1,-1,0,0,1,-2,0,0,2]\!]$$

2.20 The Mean of the Multiplication of Two Collections

Theorem 2.20.1:

Let the mean of collection $X = [\![x_1, x_2, x_3, \ldots, x_m]\!]$, be μ_X and the mean of collection $Y = [\![y_1, y_2, y_3, \ldots, y_n]\!]$, be μ_Y. Then the mean of the collection XY, is $\mu_{XY} = \mu_X \mu_Y$.

Proof:

Suppose μ_X is the mean of $X = [\![x_1, x_2, x_3, \ldots, x_m]\!]$ and μ_Y is the mean $Y = [\![y_1, y_2, y_3, \ldots, y_n]\!]$. Then

$$\mu_{XY} = \frac{\sum\limits_{i=1}^{m}\sum\limits_{j=1}^{n} x_i y_j}{mn}$$

$$\mu_{XY} = \left(\frac{\sum\limits_{i=1}^{m} x_i}{m}\right)\left(\frac{\sum\limits_{j=1}^{n} y_j}{n}\right)$$

Therefore, $\mu_{XY} = \mu_X \mu_Y$, as required.

End of proof.

Example 2.20.1

Suppose $A = [\![0, 1, 1, 2]\!]$, $C = [\![-1, 0, 0, 1]\!]$ and $F = AC$. Find μ_F.

Solution:

By Example 2.3.1, $\mu_A = 1$ and by Example 2.9.1, $\mu_C = 0$.
Now, by Theorem 2.20.1

$$\mu_F = \mu_A \mu_C$$
$$= (1)(0)$$
$$= 0.$$

Alternatively, we may calculate

$$F = AC$$
$$= [\![0, 0, 0, 0, -1, 0, 0, 1, -1, 0, 0, 1, -2, 0, 0, 2]\!]$$

by Example 2.19.1. Then, by definition we have

$$\mu_F = \dfrac{\displaystyle\sum_{i=1}^{16} f_i}{16}$$

$$= \dfrac{0}{16}$$

$$= 0.$$

2.21 The Standard Deviation of Two Collections being Multiplied Together

Theorem 2.21.1:

Let the standard deviation of collection $X = [\![x_1, x_2, x_3, \ldots, x_m]\!]$, be σ_X and the standard deviation of collection $Y = [\![y_1, y_2, y_3, \ldots, y_n]\!]$, be σ_Y . Then the standard deviation of the collection XY is $\sigma_{XY} = \sqrt{\mu_X^2 \sigma_Y^2 + \mu_Y^2 \sigma_X^2 + \sigma_X^2 \sigma_Y^2}$.

Proof:

Suppose that μ_X is the mean of $X = [\![x_1, x_2, x_3, \ldots, x_m]\!]$ and μ_Y is the mean of $Y = [\![y_1, y_2, y_3, \ldots, y_n]\!]$. Then,

$$\sigma_{XY} = \sqrt{\dfrac{\displaystyle\sum_{i=1}^{m}\sum_{j=1}^{n}\left(x_i y_j - \mu_{XY}\right)^2}{mn}}$$

$$\sigma_{XY}^2 = \dfrac{\displaystyle\sum_{i=1}^{m}\sum_{j=1}^{n}\left(x_i y_j - \mu_{XY}\right)^2}{mn}$$

$$mn\sigma_{XY}^2 = \sum_{i=1}^{m}\sum_{j=1}^{n}\left(x_i y_j - \mu_{XY}\right)^2$$

by Theorem 2.20.1. So

$$mn\sigma_{XY}^2 = \sum_{i=1}^{m}\sum_{j=1}^{n}\left(x_i y_j - \mu_X \mu_Y\right)^2$$

and

$$mn\sigma_{XY}^2 = \sum_{i=1}^{m}\sum_{j=1}^{n}\left(x_i^2 y_j^2 - 2x_i y_j \mu_X \mu_Y + \mu_X^2 \mu_Y^2\right)$$

$$mn\sigma_{XY}^2 = \left(\sum_{i=1}^{m}\sum_{j=1}^{n}x_i^2 y_j^2\right) + \left(\sum_{i=1}^{m}\sum_{j=1}^{n}\left(-2x_i y_j \mu_X \mu_Y\right)\right) + \left(\sum_{i=1}^{m}\sum_{j=1}^{n}\mu_X^2 \mu_Y^2\right)$$

48

Calculating $\sigma_X \sigma_Y$

$$\sigma_X \sigma_Y = \sqrt{\frac{\sum_{i=1}^{m}(x_i - \mu_X)^2}{m}} \sqrt{\frac{\sum_{j=1}^{n}(y_j - \mu_Y)^2}{n}}$$

We then calculate

$$\sigma_X^2 \sigma_Y^2 = \left(\frac{\sum_{i=1}^{m}(x_i - \mu_X)^2}{m}\right)\left(\frac{\sum_{j=1}^{n}(y_j - \mu_Y)^2}{n}\right)$$

$$mn\sigma_X^2 \sigma_Y^2 = \left(\sum_{i=1}^{m}(x_i - \mu_X)^2\right)\left(\sum_{j=1}^{n}(y_j - \mu_Y)^2\right)$$

$$mn\sigma_X^2 \sigma_Y^2 = \left(\sum_{i=1}^{m}(x_i^2 - 2x_i\mu_X + \mu_X^2)\right)\left(\sum_{j=1}^{n}(y_j^2 - 2y_j\mu_Y + \mu_Y^2)\right)$$

$$mn\sigma_X^2 \sigma_Y^2 = \left(\sum_{i=1}^{m}x_i^2 + \sum_{i=1}^{m}(-2x_i\mu_X) + \sum_{i=1}^{m}\mu_X^2\right)\left(\sum_{j=1}^{n}(y_j^2) + \sum_{j=1}^{n}(-2y_j\mu_Y) + \sum_{j=1}^{n}\mu_Y^2\right)$$

$$mn\sigma_X^2 \sigma_Y^2 = \begin{pmatrix} \left(\sum_{i=1}^{m}x_i^2\right)\left(\sum_{j=1}^{n}(y_j^2)\right) + \left(\sum_{i=1}^{m}(-2x_i\mu_X)\right)\left(\sum_{j=1}^{n}(y_j^2)\right) + \left(\sum_{i=1}^{m}\mu_X^2\right)\left(\sum_{j=1}^{n}(y_j^2)\right) \\ + \left(\sum_{i=1}^{m}x_i^2\right)\left(\sum_{j=1}^{n}(-2y_j\mu_Y)\right) + \left(\sum_{i=1}^{m}(-2x_i\mu_X)\right)\left(\sum_{j=1}^{n}(-2y_j\mu_Y)\right) + \left(\sum_{i=1}^{m}\mu_X^2\right)\left(\sum_{j=1}^{n}(-2y_j\mu_Y)\right) \\ + \left(\sum_{i=1}^{m}x_i^2\right)\left(\sum_{j=1}^{n}\mu_Y^2\right) + \left(\sum_{i=1}^{m}(-2x_i\mu_X)\right)\left(\sum_{j=1}^{n}\mu_Y^2\right) + \left(\sum_{i=1}^{m}\mu_X^2\right)\left(\sum_{j=1}^{n}\mu_Y^2\right) \end{pmatrix}$$

$$mn\sigma_X^2 \sigma_Y^2 = \begin{pmatrix} \left(\sum_{i=1}^{m}\sum_{j=1}^{n}x_i^2 y_j^2\right) + \left(\sum_{i=1}^{m}\sum_{j=1}^{n}(-2x_i y_j^2 \mu_X)\right) + \left(\sum_{i=1}^{m}\sum_{j=1}^{n}y_j^2 \mu_X^2\right) \\ + \left(\sum_{i=1}^{m}\sum_{j=1}^{n}-2x_i^2 y_j \mu_Y\right) + \left(\sum_{i=1}^{m}\sum_{j=1}^{n}(4x_i y_j \mu_X \mu_Y)\right) + \left(\sum_{i=1}^{m}\sum_{j=1}^{n}-2y_j \mu_X^2 \mu_Y\right) \\ + \left(\sum_{i=1}^{m}\sum_{j=1}^{n}x_i^2 \mu_Y^2\right) + \left(\sum_{i=1}^{m}\sum_{j=1}^{n}(-2x_i \mu_X \mu_Y^2)\right) + \left(\sum_{i=1}^{m}\sum_{j=1}^{n}\mu_X^2 \mu_Y^2\right) \end{pmatrix}$$

Rearranging,

$$mn\sigma_X^2\sigma_Y^2 = \left(\begin{array}{l}\left(\displaystyle\sum_{i=1}^{m}\sum_{j=1}^{n}x_i^2 y_j^2\right)+\left(\displaystyle\sum_{i=1}^{m}\sum_{j=1}^{n}\left(-2x_i y_j \mu_X \mu_Y\right)\right)+\left(\displaystyle\sum_{i=1}^{m}\sum_{j=1}^{n}\mu_X^2\mu_Y^2\right)\\[2ex] +\left(\displaystyle\sum_{i=1}^{m}\sum_{j=1}^{n}\left(-2x_i y_j^2 \mu_X\right)\right)+\left(\displaystyle\sum_{i=1}^{m}\sum_{j=1}^{n}y_j^2\mu_X^2\right)\\[2ex] +\left(\displaystyle\sum_{i=1}^{m}\sum_{j=1}^{n}-2x_i^2 y_j \mu_Y\right)+\left(\displaystyle\sum_{i=1}^{m}\sum_{j=1}^{n}\left(6x_i y_j \mu_X \mu_Y\right)\right)+\left(\displaystyle\sum_{i=1}^{m}\sum_{j=1}^{n}-2y_j\mu_X^2\mu_Y\right)\\[2ex] +\left(\displaystyle\sum_{i=1}^{m}\sum_{j=1}^{n}x_i^2\mu_Y^2\right)+\left(\displaystyle\sum_{i=1}^{m}\sum_{j=1}^{n}\left(-2x_i\mu_X\mu_Y^2\right)\right)\end{array}\right)$$

From above,

$$mn\sigma_{XY}^2 = \left(\sum_{i=1}^{m}\sum_{j=1}^{n}x_i^2 y_j^2\right)+\left(\sum_{i=1}^{m}\sum_{j=1}^{n}\left(-2x_i y_j \mu_X \mu_Y\right)\right)+\left(\sum_{i=1}^{m}\sum_{j=1}^{n}\mu_X^2\mu_Y^2\right)$$

Substituting,

$$mn\sigma_X^2\sigma_Y^2 = \left(\begin{array}{l}mn\sigma_{XY}^2\\[2ex] +\left(\displaystyle\sum_{i=1}^{m}\sum_{j=1}^{n}\left(-2x_i y_j^2 \mu_X\right)\right)+\left(\displaystyle\sum_{i=1}^{m}\sum_{j=1}^{n}y_j^2\mu_X^2\right)\\[2ex] +\left(\displaystyle\sum_{i=1}^{m}\sum_{j=1}^{n}-2x_i^2 y_j \mu_Y\right)+\left(\displaystyle\sum_{i=1}^{m}\sum_{j=1}^{n}\left(6x_i y_j \mu_X \mu_Y\right)\right)+\left(\displaystyle\sum_{i=1}^{m}\sum_{j=1}^{n}-2y_j\mu_X^2\mu_Y\right)\\[2ex] +\left(\displaystyle\sum_{i=1}^{m}\sum_{j=1}^{n}x_i^2\mu_Y^2\right)+\left(\displaystyle\sum_{i=1}^{m}\sum_{j=1}^{n}\left(-2x_i\mu_X\mu_Y^2\right)\right)\end{array}\right)$$

so,

$$mn\sigma_X^2\sigma_Y^2 - mn\sigma_{XY}^2 = \left(\begin{array}{l}\left(\displaystyle\sum_{i=1}^{m}\sum_{j=1}^{n}\left(-2x_i y_j^2 \mu_X\right)\right)+\left(\displaystyle\sum_{i=1}^{m}\sum_{j=1}^{n}y_j^2\mu_X^2\right)\\[2ex] +\left(\displaystyle\sum_{i=1}^{m}\sum_{j=1}^{n}-2x_i^2 y_j \mu_Y\right)+\left(\displaystyle\sum_{i=1}^{m}\sum_{j=1}^{n}\left(6x_i y_j \mu_X \mu_Y\right)\right)+\left(\displaystyle\sum_{i=1}^{m}\sum_{j=1}^{n}-2y_j\mu_X^2\mu_Y\right)\\[2ex] +\left(\displaystyle\sum_{i=1}^{m}\sum_{j=1}^{n}x_i^2\mu_Y^2\right)+\left(\displaystyle\sum_{i=1}^{m}\sum_{j=1}^{n}\left(-2x_i\mu_X\mu_Y^2\right)\right)\end{array}\right)$$

$$-mn\sigma_{XY}^2 = \left(\begin{array}{l}\left(\displaystyle\sum_{i=1}^{m}\sum_{j=1}^{n}\left(-2x_i y_j^2 \mu_X\right)\right)+\left(\displaystyle\sum_{i=1}^{m}\sum_{j=1}^{n}y_j^2\mu_X^2\right)\\[2ex] +\left(\displaystyle\sum_{i=1}^{m}\sum_{j=1}^{n}-2x_i^2 y_j \mu_Y\right)+\left(\displaystyle\sum_{i=1}^{m}\sum_{j=1}^{n}\left(6x_i y_j \mu_X \mu_Y\right)\right)+\left(\displaystyle\sum_{i=1}^{m}\sum_{j=1}^{n}-2y_j\mu_X^2\mu_Y\right)\\[2ex] +\left(\displaystyle\sum_{i=1}^{m}\sum_{j=1}^{n}x_i^2\mu_Y^2\right)+\left(\displaystyle\sum_{i=1}^{m}\sum_{j=1}^{n}\left(-2x_i\mu_X\mu_Y^2\right)\right)-mn\sigma_X^2\sigma_Y^2\end{array}\right)$$

50

$$mn\sigma_{XY}^2 = \left(\begin{aligned} &\left(\sum_{i=1}^{m}\sum_{j=1}^{n}\left(2x_i y_j^2 \mu_X\right) \right) - \left(\sum_{i=1}^{m}\sum_{j=1}^{n} y_j^2 \mu_X^2 \right) \\ &+ \left(\sum_{i=1}^{m}\sum_{j=1}^{n} 2x_i^2 y_j \mu_Y \right) + \left(\sum_{i=1}^{m}\sum_{j=1}^{n}\left(-6x_i y_j \mu_X \mu_Y\right) \right) + \left(\sum_{i=1}^{m}\sum_{j=1}^{n} 2y_j \mu_X^2 \mu_Y \right) \\ &- \left(\sum_{i=1}^{m}\sum_{j=1}^{n} x_i^2 \mu_Y^2 \right) + \left(\sum_{i=1}^{m}\sum_{j=1}^{n}\left(2x_i \mu_X \mu_Y^2\right) \right) + mn\sigma_X^2\sigma_Y^2 \end{aligned} \right)$$

$$mn\sigma_{XY}^2 = \left(\begin{aligned} &\left(2\mu_X \sum_{i=1}^{m} x_i \sum_{j=1}^{n} y_j^2 \right) - \left(m\mu_X^2 \sum_{j=1}^{n} y_j^2 \right) \\ &+ \left(2\mu_Y \sum_{i=1}^{m} x_i^2 \sum_{j=1}^{n} y_j \right) + \left(-6\mu_X \mu_Y \sum_{i=1}^{m}\sum_{j=1}^{n} x_i y_j \right) + \left(2m\mu_X^2 \mu_Y \sum_{j=1}^{n} y_j \right) \\ &- \left(n\mu_Y^2 \sum_{i=1}^{m} x_i^2 \right) + \left(2n\mu_X \mu_Y^2 \sum_{i=1}^{m} x_i \right) + mn\sigma_X^2\sigma_Y^2 \end{aligned} \right)$$

Now, since

$$\mu_X \mu_Y = \left(\frac{\sum_{i=1}^{m} x_i}{m} \right)\left(\frac{\sum_{j=1}^{n} y_j}{n} \right) \qquad \mu_X = \frac{\sum_{i=1}^{m} x_i}{m} \qquad \mu_Y = \frac{\sum_{j=1}^{n} y_j}{n}$$

$$mn\mu_X \mu_Y = \left(\sum_{i=1}^{m} x_i \right)\left(\sum_{j=1}^{n} y_j \right) \qquad m\mu_X = \sum_{i=1}^{m} x_i \qquad n\mu_Y = \sum_{j=1}^{n} y_j$$

We have,

$$mn\mu_X \mu_Y = \sum_{i=1}^{m}\sum_{j=1}^{n} x_i y_j$$

We use these expressions in our calculation to obtain

$$mn\sigma_{XY}^2 = \left(\begin{aligned} &\left(2\mu_X \left(m\mu_X\right) \sum_{j=1}^{n} y_j^2 \right) - \left(m\mu_X^2 \sum_{j=1}^{n} y_j^2 \right) \\ &+ \left(2\mu_Y \left(n\mu_Y\right) \sum_{i=1}^{m} x_i^2 \right) + \left(-6\mu_X \mu_Y \left(mn\mu_X \mu_Y\right) \right) + \left(2m\mu_X^2 \mu_Y \left(n\mu_Y\right) \right) \\ &- \left(n\mu_Y^2 \sum_{i=1}^{m} x_i^2 \right) + \left(2n\mu_X \mu_Y^2 \left(m\mu_X\right) \right) + mn\sigma_X^2\sigma_Y^2 \end{aligned} \right)$$

$$mn\sigma_{XY}^2 = \left(\begin{aligned} &\left(2m\mu_X^2 \sum_{j=1}^{n} y_j^2 \right) - \left(m\mu_X^2 \sum_{j=1}^{n} y_j^2 \right) \\ &+ \left(2n\mu_Y^2 \sum_{i=1}^{m} x_i^2 \right) + \left(-6mn\mu_X^2 \mu_Y^2 \right) + \left(2mn\mu_X^2 \mu_Y^2 \right) \\ &- \left(n\mu_Y^2 \sum_{i=1}^{m} x_i^2 \right) + \left(2mn\mu_X^2 \mu_Y^2 \right) + mn\sigma_X^2\sigma_Y^2 \end{aligned} \right)$$

$$mn\sigma_{XY}^2 = \left(m\mu_X^2 \sum_{j=1}^{n} y_j^2 \right) + \left(n\mu_Y^2 \sum_{i=1}^{m} x_i^2 \right) + \left(-2mn\mu_X^2\mu_Y^2 \right) + mn\sigma_X^2\sigma_Y^2$$

Now,

$$\sigma_X = \sqrt{\frac{\sum_{i=1}^{m}(x_i - \mu_X)^2}{m}} \qquad\qquad \sigma_Y = \sqrt{\frac{\sum_{j=1}^{n}(y_j - \mu_Y)^2}{n}}$$

$$\sigma_X^2 = \frac{\sum_{i=1}^{m}(x_i - \mu_X)^2}{m} \qquad\qquad \sigma_Y^2 = \frac{\sum_{j=1}^{n}(y_j - \mu_Y)^2}{n}$$

$$m\sigma_X^2 = \sum_{i=1}^{m}(x_i - \mu_X)^2 \qquad\qquad n\sigma_Y^2 = \sum_{j=1}^{n}(y_j - \mu_Y)^2$$

$$m\sigma_X^2 = \sum_{i=1}^{m}\left(x_i^2 - 2x_i\mu_X + \mu_X^2\right) \qquad\qquad n\sigma_Y^2 = \sum_{j=1}^{n}\left(y_j^2 - 2y_j\mu_Y + \mu_Y^2\right)$$

$$m\sigma_X^2 = \sum_{i=1}^{m}x_i^2 - 2\mu_X\sum_{i=1}^{m}x_i + m\mu_X^2 \qquad\qquad n\sigma_Y^2 = \sum_{j=1}^{n}y_j^2 - 2\mu_Y\sum_{j=1}^{n}y_j + n\mu_Y^2$$

From above,

$$m\mu_X = \sum_{i=1}^{m}x_i \qquad\qquad n\mu_Y = \sum_{j=1}^{n}y_j$$

Substituting,

$$m\sigma_X^2 = \sum_{i=1}^{m}x_i^2 - 2\mu_X\left(m\mu_X\right) + m\mu_X^2 \qquad\qquad n\sigma_Y^2 = \sum_{j=1}^{n}y_j^2 - 2\mu_Y\left(n\mu_Y\right) + n\mu_Y^2$$

$$m\sigma_X^2 = \sum_{i=1}^{m}x_i^2 - 2m\mu_X^2 + m\mu_X^2 \qquad\qquad n\sigma_Y^2 = \sum_{j=1}^{n}y_j^2 - 2n\mu_Y^2 + n\mu_Y^2$$

$$m\sigma_X^2 = \sum_{i=1}^{m}x_i^2 - m\mu_X^2 \qquad\qquad n\sigma_Y^2 = \sum_{j=1}^{n}y_j^2 - n\mu_Y^2$$

$$\sum_{i=1}^{m}x_i^2 = m\mu_X^2 + m\sigma_X^2 \qquad\qquad \sum_{j=1}^{n}y_j^2 = n\mu_Y^2 + n\sigma_Y^2$$

From above,

$$mn\sigma_{XY}^2 = \left(m\mu_X^2 \sum_{j=1}^{n} y_j^2 \right) + \left(n\mu_Y^2 \sum_{i=1}^{m} x_i^2 \right) + \left(-2mn\mu_X^2\mu_Y^2 \right) + mn\sigma_X^2\sigma_Y^2$$

Substituting,

$$mn\sigma_{XY}^2 = \left(m\mu_X^2 \left(n\mu_Y^2 + n\sigma_Y^2 \right) \right) + \left(n\mu_Y^2 \left(m\mu_X^2 + m\sigma_X^2 \right) \right) + \left(-2mn\mu_X^2\mu_Y^2 \right) + mn\sigma_X^2\sigma_Y^2$$

$$mn\sigma_{XY}^2 = \left(mn\mu_X^2\mu_Y^2 + mn\mu_X^2\sigma_Y^2 \right) + \left(nm\mu_Y^2\mu_X^2 + mn\mu_Y^2\sigma_X^2 \right) + \left(-2mn\mu_X^2\mu_Y^2 \right) + mn\sigma_X^2\sigma_Y^2$$

52

$$mn\sigma_{XY}^2 = mn\mu_X^2\sigma_Y^2 + mn\mu_Y^2\sigma_X^2 + mn\sigma_X^2\sigma_Y^2$$

and, after dividing both sides by mn, we obtain

$$\sigma_{XY}^2 = \mu_X^2\sigma_Y^2 + \mu_Y^2\sigma_X^2 + \sigma_X^2\sigma_Y^2$$

Taking square roots of both sides, we obtain

$$\sigma_{XY} = \sqrt{\mu_X^2\sigma_Y^2 + \mu_Y^2\sigma_X^2 + \sigma_X^2\sigma_Y^2}$$

as required.

End of proof.

Example 2.21.1

Suppose $A = [\![0, 1, 1, 2]\!]$, $C = [\![-1, 0, 0, 1]\!]$ and $F = AC$. Find σ_F.

Solution:

By Examples 2.3.1 and 2.9.1, $\mu_A = 1$ and $\mu_C = 0$.

Further, by Examples 2.4.1 and 2.9.1, $\sigma_A = \dfrac{\sqrt{2}}{2}$ and $\sigma_C = \dfrac{\sqrt{2}}{2}$.

Now,

$$\sigma_F = \sigma_{AC}$$

$$= \sqrt{\mu_A^2\sigma_C^2 + \sigma_A^2\mu_C^2 + \sigma_A^2\sigma_C^2} \quad \text{by Theorem 2.21.1}$$

$$= \sqrt{(1)\left(\frac{1}{2}\right) + \left(\frac{1}{2}\right)(0) + \left(\frac{1}{2}\right)\left(\frac{1}{2}\right)}$$

$$= \frac{\sqrt{3}}{2}.$$

Alternatively, by Examples 2.19.1 and 2.20.1, we have

$$F = AC$$
$$= [\![0, 0, 0, 0, -1, 0, 0, 1, -1, 0, 0, 1, -2, 0, 0, 2]\!]$$

and $\mu_F = 0$.

We may use the definition of the standard deviation to calculate

$$\sigma_F = \sqrt{\frac{10(0) + 4(1) + 2(4)}{16}}$$

$$= \sqrt{\frac{12}{16}}$$

$$= \frac{\sqrt{3}}{2}.$$

2.22 Definition of the Divides of a Collections

If $X = [\![x_1, x_2, x_3, \ldots, x_m]\!]$ and $Y = [\![y_1, y_2, y_3, \ldots, y_n]\!]$ were collections, and we defined $\dfrac{X}{Y}$ as

$$\frac{X}{Y} = \begin{Vmatrix} \dfrac{x_1}{y_1}, \dfrac{x_1}{y_2}, \dfrac{x_1}{y_3}, \cdots, \dfrac{x_1}{y_n}, \\[2mm] \dfrac{x_2}{y_1}, \dfrac{x_2}{y_2}, \dfrac{x_2}{y_3}, \cdots, \dfrac{x_2}{y_n} \\[2mm] \dfrac{x_3}{y_1}, \dfrac{x_3}{y_2}, \dfrac{x_3}{y_3}, \cdots, \dfrac{x_3}{y_n}, \\[2mm] \vdots \quad \vdots \quad \ddots \quad \vdots \\[2mm] \dfrac{x_m}{y_1}, \dfrac{x_m}{y_2}, \dfrac{x_m}{y_3}, \cdots, \dfrac{x_m}{y_n} \end{Vmatrix}$$ we would have several problems,

1.) The mean of $Y\left(\dfrac{X}{Y}\right)$ would not always equal the mean of X.

Proof.

Let $A = [\![2, 4]\!]$ and $B = [\![1, 2]\!]$, then

$\dfrac{A}{B} = [\![1, 2, 2, 4]\!]$ and $B\left(\dfrac{A}{B}\right) = [\![1, 2, 2, 2, 4, 4, 4, 16]\!]$

then the means $\mu_A = 3$, $\mu_B = \dfrac{3}{2}$, $\mu_{\frac{A}{B}} = \dfrac{9}{4}$ and $\mu_{B\left(\frac{A}{B}\right)} = \dfrac{35}{8}$.

So $\mu_A \neq \mu_{B\left(\frac{A}{B}\right)}$.

End of Proof.

But if we define $\dfrac{X}{Y}$ to be a collection where $\mu_X = \mu_{Y\left(\frac{X}{Y}\right)}$ and where $\sigma_X = \sigma_{Y\left(\frac{X}{Y}\right)}$ then we would not have this problem.

So I will define $\dfrac{X}{Y}$ to be a collection where $\mu_X = \mu_{Y\left(\frac{X}{Y}\right)}$ and where $\sigma_X = \sigma_{Y\left(\frac{X}{Y}\right)}$.

2.23 The Mean of the Division of Two Collections

Let $X = [\![x_1, x_2, x_3, \ldots, x_m]\!]$ and $Y = [\![y_1, y_2, y_3, \ldots, y_n]\!]$, check and make sure that $x_{Max} < 0$ or $x_{Min} > 0$ and we will defined $\dfrac{X}{Y}$ as a collection that has a mean $\mu_{\frac{X}{Y}} = \dfrac{\mu_X}{\mu_Y}$.

Proof:

$$\mu_{Y\frac{X}{Y}} = \mu_Y\left(\dfrac{\mu_X}{\mu_Y}\right)$$

$\mu_{Y\frac{X}{Y}} = \mu_X$, because $x_{Max} < 0$ or $x_{Min} > 0$

End of Proof

Example 2.23.1

Let $A = [\![2, 4]\!]$ and $B = [\![1, 2]\!]$, then $\mu_A = 3$ and $\mu_B = 1.5$

$$\mu_{\frac{A}{B}} = \dfrac{\mu_A}{\mu_B}$$

$$\mu_{\frac{A}{B}} = \dfrac{3}{1.5}$$

$$\mu_{\frac{A}{B}} = 2$$

2.24 The Standard Deviation of the Division of Two Collections

For the collection, $X = [\![x_1, x_2, x_3, \ldots, x_m]\!]$ and $Y = [\![y_1, y_2, y_3, \ldots, y_n]\!]$, check and make sure that $x_{Max} < 0$ or $x_{Min} > 0$. We will defined $\dfrac{X}{Y}$ as a collection that has a Standard Deviation of

$$\sigma_{\frac{X}{Y}} = \sqrt{\dfrac{\sigma_X^2 - \left(\dfrac{\mu_X}{\mu_Y}\right)^2 \sigma_Y^2}{\mu_Y^2 + \sigma_y^2}}.$$

Proof:

$$\sigma_{Y\frac{X}{Y}} = \sqrt{\mu_Y^2 \sigma_{\frac{X}{Y}}^2 + \mu_{\frac{X}{Y}}^2 \sigma_Y^2 + \sigma_Y^2 \sigma_{\frac{X}{Y}}^2}$$

$$\sigma_{Y\frac{X}{Y}} = \sqrt{\mu_Y^2 \left(\sqrt{\frac{\sigma_X^2 - \left(\frac{\mu_X}{\mu_Y}\right)^2 \sigma_Y^2}{\mu_Y^2 + \sigma_y^2}} \right)^2 + \mu_{\frac{X}{Y}}^2 \sigma_Y^2 + \sigma_Y^2 \left(\sqrt{\frac{\sigma_X^2 - \left(\frac{\mu_X}{\mu_Y}\right)^2 \sigma_Y^2}{\mu_Y^2 + \sigma_y^2}} \right)^2}$$

$$\sigma_{Y\frac{X}{Y}} = \sqrt{\mu_Y^2 \left(\frac{\sigma_X^2 - \left(\frac{\mu_X}{\mu_Y}\right)^2 \sigma_Y^2}{\mu_Y^2 + \sigma_y^2} \right) + \mu_{\frac{X}{Y}}^2 \sigma_Y^2 + \sigma_Y^2 \left(\frac{\sigma_X^2 - \left(\frac{\mu_X}{\mu_Y}\right)^2 \sigma_Y^2}{\mu_Y^2 + \sigma_y^2} \right)}$$

$$\sigma_{Y\frac{X}{Y}} = \sqrt{\left(\mu_Y^2 + \sigma_Y^2 \right) \left(\frac{\sigma_X^2 - \left(\frac{\mu_X}{\mu_Y}\right)^2 \sigma_Y^2}{\mu_Y^2 + \sigma_y^2} \right) + \mu_{\frac{X}{Y}}^2 \sigma_Y^2}$$

$$\sigma_{Y\frac{X}{Y}} = \sqrt{\sigma_X^2 - \left(\frac{\mu_X}{\mu_Y}\right)^2 \sigma_Y^2 + \mu_{\frac{X}{Y}}^2 \sigma_Y^2}$$

$$\mu_{\frac{X}{Y}} = \frac{\mu_X}{\mu_Y}$$

$$\sigma_{Y\frac{X}{Y}} = \sqrt{\sigma_X^2 - \mu_{\frac{X}{Y}}^2 \sigma_Y^2 + \mu_{\frac{X}{Y}}^2 \sigma_Y^2}$$

$$\sigma_{Y\frac{X}{Y}} = \sqrt{\sigma_X^2}$$

$$\sigma_{Y\frac{X}{Y}} = \sigma_X$$

End of Proof.

Example 2.24.1

Let $A = [\![2, 4]\!]$ and $B = [\![1, 2]\!]$, find $\sigma_{\frac{A}{B}}$

Then $\mu_A = 3$, $\mu_B = 1.5$, $\sigma_A = 1$, and $\sigma_B = 0.5$

$$\sigma_{\frac{A}{B}} = \sqrt{\frac{\sigma_A^2 - \left(\dfrac{\mu_A}{\mu_B}\right)^2 \sigma_B^2}{\mu_B^2 + \sigma_B^2}}$$

$$\sigma_{\frac{A}{B}} = \sqrt{\frac{(1)^2 - \left(\dfrac{3}{1.5}\right)^2 (0.5)^2}{(1.5)^2 + (0.5)^2}}$$

$$\sigma_{\frac{A}{B}} = \sqrt{\frac{0}{2.5}}$$

$$\sigma_{\frac{A}{B}} = 0$$

Example 2.24.2

Let $X = [\![2, 4, 4, 6]\!]$ and $Y = [\![4, 4, 8, 8]\!]$, find $\sigma_{\frac{X}{Y}}$

Then $\mu_X = 4$, $\mu_Y = 6$, $\sigma_X = \sqrt{2}$, and $\sigma_Y = 2$

$$\sigma_{\frac{X}{Y}} = \sqrt{\frac{\sigma_X^2 - \left(\dfrac{\mu_X}{\mu_Y}\right)^2 \sigma_Y^2}{\mu_Y^2 + \sigma_y^2}}$$

$$\sigma_{\frac{X}{Y}} = \sqrt{\frac{(\sqrt{2})^2 - \left(\dfrac{4}{6}\right)^2 (2)^2}{(6)^2 + (2)^2}}$$

$$\sigma_{\frac{X}{Y}} = \frac{\sqrt{5}}{30}$$

Chapter 3

Imprecise Numbers as Collections of Imprecise Numbers

3.1 Collection of Imprecise Numbers

A collection of imprecise numbers is a collection, $\mathbb{X} = [\![X_1, X_2, X_3, \ldots, X_m]\!]$, in which each X_i is a collection of precise numbers.

3.2 The Mean of a Collection of Imprecise Numbers

Given a collection of imprecise numbers, $\mathbb{X} = [\![X_1, X_2, X_3, \ldots, X_m]\!]$, the mean of $\mathbb{X}$ is $\mu_{\mathbb{X}} = \dfrac{\sum X_i}{m}$

.

Now try to remember how we added two imprecise numbers: we had to add each element of X to each element of Y that is, if $X = [\![x_1, x_2, x_3, \ldots, x_m]\!]$ and $Y = [\![y_1, y_2, y_3, \ldots, y_n]\!]$, to form $X + Y = [\![x_1 + y_1, x_1 + y_2, \cdots, x_1 + y_n, x_2 + y_1, x_2 + y_2, \cdots, x_2 + y_n, \cdots, x_m + y_1, x_m + y_2, \cdots, x_m + y_n]\!]$. Now try to imagine writing out what $\sum X_i$ looks like. It would be okay if $m = 2$ or 3 but what if m is large as is usually the case? It would not be practical to even try to write it out here. We need to come up with a better way of generalizing any collection. Hence, this leads me to the next chapter.

Chapter 4

The Interval in which the Imprecise Numbers Lays

4.1 Some Basics

The collection of precise numbers that defines a given imprecise number lies in some interval of the precise numbers. Usually there is an *upper limit* and a *lower limit* for the acceptable values of the imprecise number. The upper limit will be the *maximum acceptable value* for the imprecise number and is defined as follows:

In $X = [\![x_1, x_2, x_3, \ldots, x_m]\!]$, the upper limit is the value $x_{Max} \in X$ such that $x_i \leq x_{Max}$ for all $x_i \in X$.

The lower limit will be the *minimum acceptable value* for the imprecise number and is defined as follows:

In $X = [\![x_1, x_2, x_3, \ldots, x_m]\!]$, the lower limit is the value $x_{Min} \in X$ such that $x_{Min} \leq x_i$ for all $x_i \in X$.

Thus, we can form a closed interval $[x_{Min}, x_{Max}]$ containing all values of X. Sometimes, the upper and lower limits of the imprecise number are more important than the mean and standard deviation of the imprecise number, and they are also easier to work with.

4.2 What Is Meant by a Function of an Imprecise Number

Let f be a function of precise numbers, x. Then $f(x)$ will map a precise number x to another precise number, $f(x)$. Because an imprecise number is a collection of precise numbers, applying the function f to the collection X will map each precise number in the collection to another precise number, thus forming a new collection of precise numbers. More simply:

If $X = [\![x_1, x_2, x_3, \ldots, x_m]\!]$, then $f(X) = [\![f(x_1), f(x_2), f(x_3), \ldots, f(x_m)]\!]$.

Because $f(X)$ is a collection of precise numbers, $f(X)$ is also an imprecise number and it will have an upper and lower limit. The maximum of any collection X can be found by using the function $\max(X)$, defined as $\max(X) = x_{Max}$. Likewise, the minimum of any collection X can be found by using the function $\min(X)$ which is defined as $\min(X) = x_{Min}$.

4.3 An Increasing Function, $f(x)$

A function for which the value of $f(x)$ either remains the same or increases as x increases is called an increasing function. The following is an example of an increasing function, $f(x)$.

Example of an increasing function

A function $f(x)$ is increasing on the closed interval $\left[\min(A), \max(A)\right]$ if, for all $x_1, x_2 \in \left[\min(A), \max(A)\right]$, we have $f(x_1) \leq f(x_2)$ whenever $x_1 \leq x_2$.

Theorem 4.3.1:

If $f(x)$ is an increasing function on the collection A, and $x \in \left[\min(A), \max(A)\right]$, then $\min(f(x)) = f(\min(A))$ and $\max(f(x)) = f(\max(A))$. This means that all values of $f(A)$ will be contained in the closed interval, $\left[f(\min(A)), f(\max(A))\right]$.

Proof:

Suppose that $f(x)$ is increasing on $\left[\min(A), \max(A)\right]$.

Let $x \in \left[\min(A), \max(A)\right]$. Then $\min(A) \leq x \leq \max(A)$ and, since $f(x)$ is increasing,
$$f(\min(A)) \leq f(x) \leq f(\max(A)).$$

Therefore, all values of $f(x)$ are contained in the closed interval, $\left[f(\min(A)), f(\max(A))\right]$, and our result follows.

End of proof.

4.4 A Decreasing Function $f(x)$

A decreasing function is one for which the value of $f(x)$ either remains the same or decreases, but never increases, as x increases. The following is an example of a decreasing function.

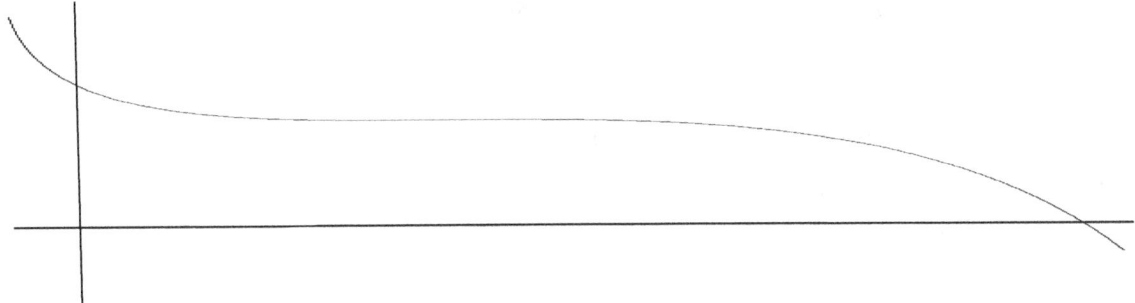

Example of a decreasing function

A function, $f(x)$, is decreasing on the closed interval $\left[\min(A),\max(A)\right]$ if, for all $x_1, x_2 \in \left[\min(A),\max(A)\right]$, we have $f(x_1) \geq f(x_2)$ whenever $x_1 \leq x_2$.

Theorem 4.4.1:

If $f(x)$ is a decreasing function on the collection A, and $x \in \left[\min(A),\max(A)\right]$, then $\min(f(x)) = f(\max(A))$ and $\max(f(x)) = f(\min(A))$. Note that the minimum and maximum occur at the opposite ends of the interval to those for an increasing function. This means that $f(A)$ will be contained in the closed interval, $\left[f(\max(A)), f(\min(A))\right]$.

This theorem can be proved in an analogous way to Theorem 4.3.1.

4.5 Definition of a Function of Two Collections, $f(A,B)$

In the earlier sections, we had learned how to compute the maximum and minimum values for functions of one collection when the function was increasing or decreasing. Now we will learn what happens to functions of two collections. The following are some examples of such functions: $A+B$, $A-B$, $-A+B$, $-A-B$, AB, $\dfrac{A}{B}$, etc.

If $A = [\![a_1, a_2, a_3, \ldots, a_m]\!]$ and $B = [\![b_1, b_2, b_3, \ldots, b_n]\!]$, then $f(A, B)$ will be a new collection constructed from all of the possible combinations of A and B as follows:

$$f(A, B) = [\![f(a_1, b_1), f(a_1, b_2), \ldots, f(a_1, b_n), f(a_2, b_1), f(a_2, b_2), \ldots, f(a_2, b_n), \ldots, f(a_m, b_n)]\!]$$

If the first collection has m elements and the second collection has n elements, then the function of the two collections will have mn elements.

Suppose an imprecise number, A contains elements $a_1, a_2, a_3 \in A$, such that $a_1 \le a_2 \le a_3$ and an imprecise number, B contains elements $b_1, b_2, b_3 \in B$, such that $b_1 \le b_2 \le b_3$. Then any ordering of the values of $f(a_1, b_1)$, $f(a_2, b_2)$ and $f(a_3, b_3)$ is possible. The diagram below shows the situation where $f(a_1, b_1) \le f(a_2, b_2) \le f(a_3, b_3)$, but this is not always the case.

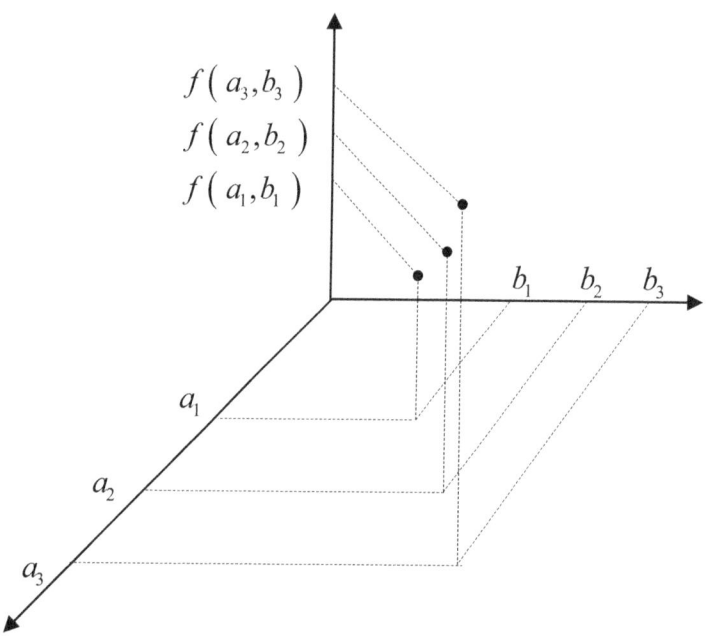

Now, imagine if we wanted to plot points for all of the possible values of $f(A, B)$. If A has m elements, and B has n elements, then there would be mn of these points to plot. It would be nice if we could find the values of $\min(f(A, B))$ and $\max(f(A, B))$, because then all the values of $f(A, B)$ would lie in the closed interval $\left[\min(f(A, B)), \max(f(A, B))\right]$. It would also be nice to find conditions on $f(x, y)$ under which all the values of $f(A, B)$ would lie in the closed interval, $\left[\min(f(A, B)), \max(f(A, B))\right]$. We will explore some of these conditions in sections 4.6 to 4.10 of this chapter.

4.6 $f(x,y)$ is Increasing for both x and y

The function $f(x,y)$ is increasing for both x and y if both of the following conditions are satisfied:

First, $f(x,y)$ is increasing for x alone. This means that, if we fix y, then $f(x,y)$ becomes an increasing function of x.

Second, $f(x,y)$ is an increasing function for y alone. This means that, if we fix x, then $f(x,y)$ becomes an increasing function of y.

4.7 Definition of a Function $f(x,y)$ which is Increasing in Both x and y

If $x \in [x_{Min}, x_{Max}]$, where $x_{Min} \neq x_{Max}$, and $y \in [y_{Min}, y_{Max}]$, where $y_{Min} \neq y_{Max}$, then $f(x,y)$ is increasing in both x and y if the following conditions are satisfied:

Condition 1: If $x_i, x_j \in [x_{Min}, x_{Max}]$ and $y_r \in [y_{Min}, y_{Max}]$, then $f(x_i, y_r) \leq f(x_j, y_r)$ whenever $x_i \leq x_j$.

Condition 2: If $x_s \in [x_{Min}, x_{Max}]$ and $y_k, y_l \in [y_{Min}, y_{Max}]$, then $f(x_s, y_k) \leq f(x_s, y_l)$ whenever $y_k \leq y_l$.

4.8 When $f(A,B)$ if $f(x,y)$ is Increasing in Both x and y

Theorem 4.8.1:

Let A and B be collections of precise numbers and suppose that $f(x,y)$ is increasing in x on $[\min(A), \max(A)]$ and increasing in y on $[\min(B), \max(B)]$. Then
$\min(f(x,y)) = f(\min(A), \min(B))$ and $\max(f(x,y)) = f(\max(A), \max(B))$.
Moreover, if $c \in f(A,B)$, then we must have $c \in [f(\min(A), \min(B)), f(\max(A), \max(B))]$.

The following diagram shows an example of a function that is increasing in both x and y. In the diagram, $a_1 \le a_2 \le a_3$, $b_1 \le b_2 \le b_3$, $a_1 \le x \le a_3$, and $b_1 \le y \le b_3$.

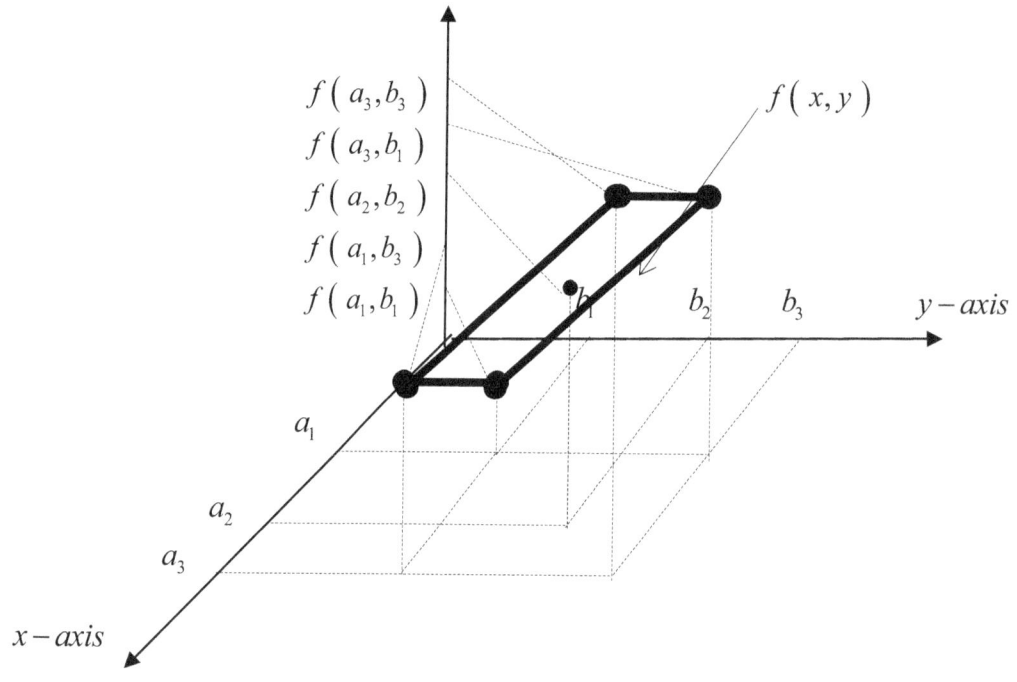

Proof of Theorem 4.8.1:

Let $x_{\min} = \min(A)$, $x_{\max} = \max(A)$, $y_{\min} = \min(B)$, and $y_{\max} = \max(B)$.

Suppose, $f(x,y)$ is increasing in x on $[x_{Min}, x_{Max}]$, where $x_{Min} \ne x_{Max}$ and is increasing in y on $[y_{Min}, y_{Max}]$, where $y_{Min} \ne y_{Max}$.

Let $x_i, x_j \in [x_{Min}, x_{Max}]$ and $y_k, y_l \in [y_{Min}, y_{Max}]$ satisfy $x_{Min} \le x_i \le x_j \le x_{Max}$ and $y_{Min} \le y_k \le y_l \le y_{Max}$.

I will show all 16 combinations of x and y are bounded by (x_{Min}, y_{Min}) and (x_{Max}, y_{Max}).

Then, $f(x_{Min}, y_{Min}) \le f(x_i, y_{Min}) \le f(x_i, y_k) \le f(x_j, y_k) \le f(x_{Max}, y_k) \le f(x_{Max}, x_l) \le f(x_{Max}, y_{Max})$ and

$f(x_{Min}, y_{Min}) \le f(x_i, y_{Min}) \le f(x_i, y_l) \le f(x_j, y_l) \le f(x_{Max}, y_l) \le f(x_{Max}, y_{Max})$ and

$(x_{Min}, y_{Min}) \le (x_i, y_{Max}) \le (x_j, y_{Max}) \le (x_{Max}, y_{Max})$ and

$(x_{Min}, y_{Min}) \le (x_{Min}, y_k) \le (x_{Min}, y_l) \le (x_{Max}, y_l) \le (x_{Max}, y_{Max})$.

It follows that $\min(f(x,y)) = f(\min(A), \min(B))$ and

$\max(f(x,y)) = f(\max(A), \max(B))$.

Thus, if $c \in f(A, B)$, then $c \in \left[f(\min(A), \min(B)), f(\max(A), \max(B)) \right]$, as required.

End of proof.

4.9 Definition of a Function $f(x,y)$ which is Increasing for x and Decreasing for y

If $x \in [x_{Min}, x_{Max}]$, where $x_{Min} \neq x_{Max}$ and $y \in [y_{Min}, y_{Max}]$, where $y_{Min} \neq y_{Max}$, then $f(x,y)$ is increasing for x and decreasing for y if the following conditions are satisfied:

Condition 1: If $x_i, x_j \in [x_{Min}, x_{Max}]$ and $y_r \in [y_{Min}, y_{Max}]$, then $f(x_i, y_r) \leq f(x_j, y_r)$, whenever $x_i \leq x_j$.

Condition 2: If $x_s \in [x_{Min}, x_{Max}]$ and $y_k, y_l \in [y_{Min}, y_{Max}]$, then $f(x_s, y_k) \geq f(x_s, y_l)$, whenever $y_k \leq y_l$.

Theorem 4.9.1:

Suppose that A and B are collections of precise numbers and that $f(x,y)$ is increasing in x on $[\min(A), \max(A)]$ and decreasing in y on $[\min(B), \max(B)]$. Then $\min(f(x,y)) = f(\min(A), \max(B))$ and $\max(f(x,y)) = f(\max(A), \min(B))$. Moreover, if $c \in f(A,B)$, then we must have

$$c \in [f(\min(A), \max(B)), f(\max(A), \min(B))].$$

The proof of Theorem 4.8.1 is analogous to the proof of Theorem 4.7.1.

4.10 Definition of a Function $f(x,y)$ which is Decreasing for x and Increasing for y

If $x \in [x_{Min}, x_{Max}]$, where $x_{Min} \neq x_{Max}$ and $y \in [y_{Min}, y_{Max}]$, where $y_{Min} \neq y_{Max}$, then $f(x,y)$ is decreasing for x and increasing for y if the following conditions are satisfied:

Condition 1: If $x_i, x_j \in [x_{Min}, x_{Max}]$ and $y_r \in [y_{Min}, y_{Max}]$, then $f(x_i, y_r) \geq f(x_j, y_r)$, whenever $x_i \leq x_j$.

Condition 2: If $x_s \in [x_{Min}, x_{Max}]$ and $y_k, y_l \in [y_{Min}, y_{Max}]$, then $f(x_s, y_k) \leq f(x_s, y_l)$, whenever $y_k \leq y_l$.

4.10 When $f(A,B)$ if $f(x,y)$ is Decreasing in x and is Increasing in y

Theorem 4.10.1:

Suppose that A and B are collections of precise numbers and that $f(x,y)$ is decreasing in x on $\left[\min(A),\max(A)\right]$ and increasing in y on $\left[\min(B),\max(B)\right]$. Then $\min(f(x,y))=f(\max(A),\min(B))$ and $\max(f(x,y))=f(\min(A),\max(B))$. Moreover, if $c \in f(A,B)$, then we must have
$$c \in \left[f(\max(A),\min(B)), f(\min(A),\max(B))\right].$$

The proof of Theorem 4.9.1 is analogous to the proof of Theorem 4.7.1.

4.11 Definition of a Function $f(x,y)$ which is Decreasing in Both x and y

If $x \in \left[x_{Min}, x_{Max}\right]$, where $x_{Min} \neq x_{Max}$ and $y \in \left[y_{Min}, y_{Max}\right]$ where, $y_{Min} \neq y_{Max}$, then $f(x,y)$ is decreasing for both x and y if the following conditions are satisfied:

Condition 1: If $x_i, x_j \in \left[x_{Min}, x_{Max}\right]$ and $y_r \in \left[y_{Min}, y_{Max}\right]$, then $f(x_i, y_r) \geq f(x_j, y_r)$, whenever $x_i \leq x_j$.

Condition 2: If $x_s \in \left[x_{Min}, x_{Max}\right]$ and $y_k, y_l \in \left[y_{Min}, y_{Max}\right]$, then $f(x_s, y_k) \geq f(x_s, y_l)$, whenever $y_k \leq y_l$.

4.11 When $f(A,B)$ if $f(x,y)$ is Decreasing in Both x and y

Theorem 4.11.1:

Suppose that A and B are collections of precise numbers and that $f(x,y)$ is decreasing in x on $\left[\min(A),\max(A)\right]$ and decreasing in y on $\left[\min(B),\max(B)\right]$. Then $\min(f(x,y))=f(\max(A),\max(B))$ and $\max(f(x,y))=f(\min(A),\min(B))$. Moreover, if $c \in f(A,B)$, then we must have
$$c \in \left[f(\max(A),\max(B)), f(\min(A),\min(B))\right].$$

The proof of Theorem 4.11.1 is analogous to the proof of Theorem 4.7.1.